Engineering Safe and Secure Software Systems

For a complete listing of titles in the
Artech House Computer Security Series,
turn to the back of this book.

Engineering Safe and Secure Software Systems

C. Warren Axelrod

ARTECH HOUSE

BOSTON | LONDON
artechhouse.com

Library of Congress Cataloging-in-Publication Data
A catalog record for this book is available from the U.S. Library of Congress.

British Library Cataloguing in Publication Data
A catalogue record for this book is available from the British Library.

Cover design by Vicki Kane

ISBN 13: 978-1-60807-472-3

© 2013 ARTECH HOUSE
685 Canton Street
Norwood, MA 02062

10 9 8 7 6 5 4 3 2 1

To Judy, David, Nicole, Elisabeth, Evan, and Jolie,
with wishes for a safer and more secure world for future generations

Contents

Preface

The best laid plans o' Mice an' Men, Gang oft agley ...
—Robert Burns, *To a Mouse*

The initial concept for this book arose some 3 to 4 years ago. However, it was quite different from how the book turned out. I had spent much of the previous decade working on application security, particularly software assurance. Several years ago, I had the good fortune of being the technical lead on a software assurance initiative for the banking and finance sector, supported by the Financial Services Technology Consortium. The first phase of the project provided insights from thought leaders from independent software vendors (ISVs), information security tools and services vendors, industry and professional associations, academia, and a number of leading financial institutions. A collection of state-of-the-art practices was assembled, with the intention of using the "best" approaches for assuring the quality of software through industry-sponsored testing. This work provided a substantial amount of the research that was behind the BITS publication *Software Assurance Framework* [1].

The ultimate goal of the software assurance initiative was to establish a state-of-the-art testing facility for the financial services industry using methods, tools, and services chosen from the initial research. Unfortunately, the financial meltdown of 2008 interceded. Various mergers and restructurings took place, so that attention was turned to other more pressing matters.

The financial services industry and other critical infrastructure sectors continue to be in dire need of such testing laboratories to ensure that commonly used software meets agreed-upon security standards. Others, such as Joel Brenner [2], also espouse such a concept as some form of Consumers-Union-like autonomous testing service. But who would support and fund such an effort?

Perhaps an interesting indicator as to how such testing laboratories might be created is the example of Huawei, the Chinese telecommunications giant, which built the Cyber Security Evaluation Centre in England. The Centre's purpose is to assure potential British customers of Huawei products and services that the company's technology can be trusted not only to do what customers want them to do (and nothing more) but also that Huawei's products cannot be successfully attacked by cyber criminals, terrorists or foreign spies, as reported in *The Economist* [3].

My initial book concept was to cover software assurance with particular emphasis on requirements, context, and the like, with some, but relatively little mention of the differences between developing security-critical and safety-critical software systems, and how to bring these diverse fields together. Then what caused me to put so much more emphasis on the engineering of the security and safety of software systems? It was a realization that professionals in the software security and safety fields have minimal interaction and are generally not familiar with the other engineers' knowledge base and standard procedures. This argument is supported from personal experience as well as from the observation that the majority of books and articles covering one area have little if any mention of the other field. I first became aware of the difference in culture and approach when I did some work with a small government contractor that builds safety-critical software systems. Their approach was so different from what I had been used to throughout an IT career in financial services. The ratio of programming to testing appeared to be reversed for each type of system, with security-critical information systems receiving much less testing relative to the extensive verification and validation processes to which safety-critical control systems are subjected.

A number of other factors began pointing to the need for developing a book that would help security software engineers understand safety and vice versa. However, the most influential event was a serendipitous conversation at the *2010 IEEE International Conference on Homeland Security Technology* in Waltham, Massachusetts with Peter Gutgarts. He and Aaron Termin were presenting on their topic "Security-Critical versus Safety-Critical Software." Peter Gutgarts and I discussed the subject at length and what I gained from the conversation was the realization that so little attention was being paid to combined safety and security aspects of critical software systems. These systems, which include so-called cyber-physical systems, are rapidly becoming crucial to the future of modern economies. Before our eyes, software systems are being developed that merge sophisticated modern web applications with traditional industrial control systems. Such cyber-physical systems offer the best—and the worst—of both worlds. For example, the ability to monitor and control electricity meters remotely creates many new features, such as determining and reporting unusual activities or cessation of service, but they also expose the electricity grid to hackers with malevolent and destructive intentions.

Today, the security of the smart grid and analogous systems for water and natural gas distribution and the like are being discussed and are a clear concern of both government and the private sector. Yet, beyond the talk, what is actually being done is woefully inadequate. Is this a case of the road to hell being paved with good intentions? It seems that there is agreement in principle as to the need, yet the lack of subject-matter expertise and processes for coordination and collaboration are thwarting attempts to make it all happen. We have not yet built the necessary bridges between security professionals and safety engineers to enable fully coordinated efforts. The cultural differences and the goals of the security and safety silos are such that they do not interact and do not understand the needs of the other group.

Seeing the lack of interaction, in the field and in the literature, I determined that the software engineering field needs to initiate conversations that will reduce and eliminate the lack of communication between information security professionals and software safety engineers.

Brenner [2] describes how, to their credit, the U.S. military found a way to ensure that the efforts of diverse experts would be coordinated and consistent across Defense Departments by creating the Joints Chiefs of Staff. As Brenner puts it:

"Joint organization has paid extraordinary operational dividends but it was a struggle to make it work. Overcoming intense service-level loyalties took time... The [1986 Goldwater-Nichols Act] is one of the most important organizational reforms in the history of the United State government..."

If the military can do it, so perhaps could the rest of government and the private sector when it comes to coordinating the efforts to ensure that critical software systems are both secure and safe. The need is there for solutions that will overcome the security, safety, and other deficiencies in the software development processes and the software systems that they produce.

The original direction of improving the software assurance processes and building testing facilities continues to be extremely important and needs to be addressed vigorously in short order. However, I chose to focus this book on reducing the huge gap that exists between the software safety and security engineering fields. No doubt others will revisit my original software-assurance topics at a later date. For now, I will focus on what I see to be one of the most pressing needs of today's software engineering world—the collaboration between software security and safety engineers in the creation of complex systems of systems, particularly cyber-physical systems.

While I have covered many concepts in the software systems safety-security space, this work should not be considered an end in itself. It provides a starting point from which you can examine further the many complex topics that have been discussed in this book. Efforts on your part to expand upon that which you read here may be challenging—but they will also be exciting. You will be rewarded with the knowledge that you are learning more about what is surely one of the most critical areas of software engineering. Good luck with your endeavors.

Endnotes

[1]   BITS Division of the Financial Services Roundtable, *Software Assurance Framework*, January 2012. Available at http://www.bits.org/publications/security/BITSSoftwareAssurance0112.pdf. Accessed August 26, 2012.

[2]   Brenner, J., *America the Vulnerable: Inside the New Threat Matrix of Digital Espionage, Crime, and Warfare*, New York: Penguin Press, 2011.

[3]   "Briefing Huawei: The Company that Spooked the World," *The Economist*, August 4–10, 2012, pp.19–23.

Foreword

Computer security was once a casual, even polite, discipline that was practiced in business and government via little tips: "Don't use your name as a password," published policy rules: "Please inform the security team of any Internet-facing servers," and off-the-shelf products: "Based on our scan, your system is free of malware!"

This era of casualness has unfortunately long since passed, and technology practitioners, ranging from system administrators, to software engineers, to casual users, all understand that malicious threats to organizational assets are real. Just about every major organization on the planet has been hit with advanced persistent threats (APTs), distributed denial of service (DDOS) threats, or both—and the effects are often lethal. Safety-critical issues have been similarly intense with often serious consequences for the organization.

Warren Axelrod's new book is an important contribution to the disciplines of security and safety—offering information that will be vital to anyone connected in any way to the protection of information and assets. While his book targets the software engineering life cycle, many of its security and safety points can be expanded and extrapolated to more general contexts. The discussions and examples are practical and provide considerable insights for professionals interested in immediate-term benefit.

The material on safeguarding public data and intellectual property is particularly useful, given so many recent public issues. Its emphasis on risk, and how it relates to security and safety-critical systems is also particularly relevant in the current global environment.

Newly published works on this topic of computer security and the related topic of system safety seem to appear on a regular basis, perhaps even daily. But few are produced by someone with the experience and knowledge of this capable author, a longtime friend of mine. His background and expertise are unique and I am pleased that he's taken the time to share a portion of his knowledge on protecting assets.

If you have already purchased this book, then by all means, please turn the page and begin reading. If, however, you are flipping through these pages online (or at your bookstore), then I strongly suggest that you add this book to your cart. It's worth the read.

Edward G. Amoroso[1]
AT&T Chief Security Officer and Senior Vice President

---

1. Dr. Edward G. Amoroso is responsible for all areas of information security including planning, design, development, implementation, and operations support for AT&T's extensive global network and communications. Ed Amoroso is an adjunct professor of computer science at Stevens Institute of Technology and holds a Ph.D. and masters degree in computer science. He is a frequent expert speaker before Senate subcommittees on the topic of cyber security.

1

Introduction

A paradigm shift is underway, and a number of recent threads point towards a fusion of security with software engineering, or at the very least to an influx of software engineering ideas.
—Ross J. Anderson, Cambridge University, *Why Cryptosystems Fail,*
   1993 [1]

While the general concept of safety and reliability is understood by most parties, the specialty of software safety and reliability is not.
—Debra S. Hermann, *Software Safety and Reliability* [2]

Preamble

Despite heroic efforts by security and safety software engineers, information and physical security professionals, and those involved with creating and deploying safe and secure software, software systems still succumb to willful and accidental attacks and are unable to avoid completely malicious or unintended damage to human life and the environment.

One reason might be the shortage of subject-matter experts with in-depth knowledge and broad experience in application security, software system safety, software assurance, and related subjects. Even so, we must question why we aren't doing a better job of building, operating, and maintaining safe and secure software-intensive systems. After all, there is a considerable body of knowledge, much of which will be discussed in this book, about building, deploying, operating, and maintaining safety-critical and security-critical software-intensive systems.

These problems do not lie with lack of knowledge or ability on the part of experts; rather, they have to do with the omission of critical elements—not the

1

least of which is communication among diverse stakeholders—which is needed for securing software systems and making them safe. Recommendations by researchers, particularly as they relate to information security and system safety, are often tentative and normative in nature and seldom usable in the real world. In some cases, this is due to the lack of authority and power of those asserting the need for action to improve the safety and security of software systems. In other instances, researchers are unrealistic about the application of certain techniques, recommending activities that may be too difficult or expensive for the average development shop to implement.

There is also a mind-numbing, imprecise, and often misleading vocabulary that results from the art of creating safe and secure software systems sitting at the intersection of so many different fields with stakeholders who do not communicate properly with one another. As a result, conversations and interactions among safety and security experts are frequently obscure and misguided, and mutual agreement about the relative importance of security and safety for specific critical systems is seldom reached. This can produce systems where weaknesses of one segment are detrimental to otherwise high-integrity, high-assurance systems. This failure to communicate leads to critical systems that are vulnerable to attack and where successful attacks can result in considerable loss of assets or harm to life and the environment.

In this book, we shall look beyond the standard texts, although they will be referenced and studied extensively. Rather, we shall base our inquiries on the simple question: "What will it take to ensure that the software systems, upon which our economic and physical lives depend, are trustworthy, dependable, and sustainable?"

Part of the answer will come from a rigorous examination of the terms and definitions that we use. For example, the words "security" and "safety" are commonly used interchangeably, but if examined more closely, their differences are the key to understanding how to create secure, safe, and dependable software-intensive systems. Then we have the lack of rigorous testing, needed to ensure that systems will not misbehave or allow manipulation that could lead to disastrous outcomes.

It should be noted that even today there is little communication between security and safety silos involved in building complex security-critical and safety-critical software intensive systems. This isolation leads to application software without reference to the platforms and infrastructures upon which it operates.

Context is a key aspect of achieving security, safety, reliability, availability, integrity, resiliency, and the like. A system has a very different set of demands if it is exposed to public communications networks such as the Internet, than if it is isolated from external networks; however, there are ways to bridge such an "air gap," as was demonstrated by the Stuxnet attack on the Iranian uranium processing plants.

Furthermore, the meaningfulness of application security metrics must be examined. Measurements that we have and use are often not up to the job, as witnessed by the frequent injection of malware, submission to denial of service attacks, and common system failures. To what extent is it a question of not using the right metrics versus not collecting the data upon which to base the metrics in the first place? The gathering and analysis of appropriate metrics is the basis of risk analysis. If we don't know what is going on inside of our software systems, then we cannot determine the level of risk being incurred, and consequently, we are then unable to mitigate those risks effectively.

Additionally, we have supply chain issues. Can we trust commercial software generally and software developed offshore in particular? Do we know whose fingers touched the software (and the hardware on which it runs) during the development life cycle, deployment, and maintenance? Is there reason to suspect that those with evil intentions have inserted back doors and malware into the software systems? How can we ensure not only the security and integrity of applications in general but also the specific copy that we are using?

This book addresses what the author considers to be real issues and impediments to the development and operation of safe and secure mission-critical, software-intensive systems. It does not ignore the common vulnerabilities and errors found in applications, such as those listed in The Open Web Application Security Project (OWASP) top ten or The System Administration, Networking, and Security Institute (SANS)/Common Weakness Enumeration (CWE) top twenty and described in Appendix A. Rather, it takes the view that these specific risks are important and must necessarily be dealt with, but they alone will not solve the problem of deficient software systems. As we will discover, merely solving these problems is not sufficient to achieve a level of assurance that critical software systems meet high standards of security and safety. What is needed is a pragmatic view of what it takes to achieve the optimal levels of security and safety, however optimality is defined. It is the mission of this book to focus on the neglected aspects of software systems engineering that stand the chance of our making a material improvement in the state of software, a state that currently is deteriorating rapidly.

Scope and Structure of the Book

As mentioned previously in this book, it was to be a treatise on software assurance. However, it rapidly became apparent that, in order to come up with an approach that would cover both the security and safety of software systems, it was necessary to go back in time to see how software engineering has evolved over recent decades. We also sought to learn why there is so little overlap between the areas of interest of software security engineers and software safety engineers. Books on software security seldom include references to software safety, and vice versa; it is as if each party were totally oblivious as to what goes on with their software engineering counterparts in the other area. This book attempts to bridge that gap between cyber security and physical safety professionals. This approach recognizes that these so-called cyber-physical systems evolve over time and, as they do, there will be an increasing need for those who understand both security and safety to interrelate.

The first half of this book, Chapters 2 through 5, examines the history and evolution of systems engineering, covering software systems and the security and safety attributes of those systems. The next section, Chapter 6 and 7, addresses risk issues and metrics, which are fundamental to the assessment and management of systems in general and secure and safe software systems in particular. In the next sections, Chapters 8, 9, and 10, we describe the processes available for developing software systems, with particular emphasis on secure and safe software systems. Chapter 11 discusses how economic and behavioral factors influence the degree to which stakeholders participate in and support security and safety features and attributes.

Appendix A provides lists and explanations of programming weakness and errors as presented by OWASP, SANS, and others. Appendix B compares the ISO/IEC 12207 and CMMI®-DEV process areas. Appendix C provides checklists in support of security-critical systems, and Appendix D does the same for safety-critical systems. These latter two sets of checklists serve two purposes. The first is to help software engineers ensure that they have considered a full, though not complete, set of issues relating to their area of specialty (namely security or safety). The second purpose is to enable software security engineers to learn what factors make for safe systems, and to help software safety engineers determine which issues are important for building and operating security-critical systems.

Acknowledgments

While I have been around computer systems throughout my career, my interest in application security in particular came from discussions in the mid-1990s with Tom Whitman at Pershing LLC, where I was the chief information security officer. Tom was a keen advocate of including security signoffs throughout the software development life cycle and strongly supported my efforts to introduce a role for security in the life cycle process. Around that time, very few security professionals were giving much attention to building-in security as software development passed through its sequence of phases. As well, there were few experts in the area. However, I was fortunate enough to have Ken van Wyk train some of Pershing's software development staff in how to design and code programs using secure methods. We also held discussions with Dr. Gary McGraw of Cigital, who is recognized as a one of the top thought leaders in the field.

In 2009–2010, I led a Software Assurance Initiative (SAI) project for Financial Services Technology Consortium (FSTC), which is now part of BITS/FS Roundtable. This gave me an opportunity to work with Dr. Daniel Schutzer and Roger Lang. Project team members included representatives from financial institutions, professional associations, software vendors, software security service providers, and academia. The purpose of the SAI project was to pull together state-of-the-art practices in application security for use by financial firms. Team members included pioneers in their respective specialties, and everyone benefited from the outstanding presentations that they gave. I wish to thank team participants for greatly enhancing my personal knowledge of the subject. The ultimate goal was to set up an industry software test center in which commonly used software products could be evaluated. The industry was not ready for the test lab at the time, although the concept behind it (namely testing and certifying software products for a group rather than by individual firm to save on the cost of testing) is still valid, and was mentioned in Joel Brenner's recent book [3].

Dr. Jennifer L. Bayuk, who is the program director for cyber security at the Stevens Institute of Technology, and was formerly an eminent information security executive for a large financial services firm, introduced me to leading-edge software engineering as taught at Stevens.

Thanks are also due to the staff at Decilog Inc., particularly Neal and Scott Marchesano, Bruce Hennessy, and Jeff Valino, who enabled me to learn so much more about how safety-critical, software-intensive control systems are developed and tested. In addition, I owe a great deal to Artech House editors Deidre Byrne, Samantha Ronan, and Judi Stone, who kept the pressure on until I finally delivered the manuscript. I really don't think I could have done it without their support and urging. Also, my anonymous reviewer was extremely helpful in pointing out omissions and errors, and in making some very helpful suggestions.

Finally, I want to thank my family, especially my wife, Judy, for putting up with so much during the writing of this book. It really did interfere with so many other things that we wanted to do together. Unfortunately, that seems to be the price that most authors and their families are required to pay.

Endnotes

[1] See http://web.cs.wpi.edu/~guttman/cs559_website/wcf.pdf, last accessed on July 23, 2012.

[2]   Herrmann, D. S., *Software Safety and Reliability*, Los Alamitos, CA: IEEE Computer Society, 1999.

[3]   Brenner, J., America the Vulnerable: Inside the New Threat Matrix of Digital Espionage, Crime, and Warfare, New York: Penguin Press, 2011.

2

Engineering Systems

There is nothing more difficult to carry out, nor more doubtful of success, nor more dangerous to handle, than to initiate a new order of things [or create a new system]. For the reformer has enemies in all those who profit by the old order, and only lukewarm defenders in all those who would profit by the new order, this lukewarmness arising partly from fear of their adversaries, who have the laws in their favor; and partly from the incredulity of mankind, who do not truly believe in anything new until they have had the actual experience of it.
—Niccolo Machiavelli, *The Prince,* 1513

You have to be run by ideas, not hierarchy. The best ideas have to win.
—Steve Jobs, Cofounder, Apple Computer

Author's note: Steve Jobs's remark raises questions as to the place of creativity and innovation in the systems engineering process, particularly the engineering of safe and secure software systems. One might presume that there can be little opportunity for creativity and flexibility in the clearly-defined, highly-structured world of making software systems both safe and secure. However, it can be argued that the essence of the problems that we face in software engineering is exactly this; that is, not enough imagination and innovation are being brought to bear on the traditional systematic engineering processes that produce today's software systems. Perhaps we lack the out-of-box thinking that is needed to anticipate the range of threats to which our systems are subjected and to come up with innovative approaches to avoidance, deterrence, and remediation. In this book, we attempt to bring some measure of insight into the somewhat moribund approaches that, to date, may have lacked sufficient innovative and effective ideas. First we will look at what exists today and what is missing. We will then suggest how we might alter the balance between good and evil so that the good guys have the better ideas…and win.

Introduction

Before embarking on our journey through the maze that is software systems engineering and into the land of safety and security engineering (as these apply to software systems), we will dabble in the more general field of systems engineering, which is by far better established and has higher credibility than software systems security and safety engineering.

It is quite remarkable that, with a foundation as sound and effective as systems engineering has become over the past 70 or so years, the building and deployment of safe, secure, dependable, resilient, and reliable software systems remains so inadequate. It is not that the fundamental principles underlying the engineering of software systems are particularly lacking, but clearly software systems engineering has not yet received the attention and support that are needed to establish the pursuit of safe and secure systems as a respected field in its own right.

Before diving into the engineering of software systems, we will first examine the broader field of systems engineering—its approaches, practices, standards, and so on. The intent is to provide a foundation upon which to build lower-level approaches relating to software safety and security, as well as to help identify and resolve the gaps between systems engineering and software engineering. By carrying over some of the structure and processes of the longer-established discipline of systems engineering, we can apply the experience and proven practices of systems engineering to the less mature art of engineering safe and secure software systems.

Some Initial Observations

Later in this chapter, we will work with a specific hierarchy of terms in order to facilitate more meaningful definitions that we will use throughout the book. The hierarchy trickles down from general systems to software-oriented and hardware-oriented systems, and then to the security and safety of those systems. While the focus of this book is on software systems, there is increasing interest in the need for codesigning applications and related systems, particularly for high-performance computing. We will, therefore, consider the implications of the need to design applications in conjunction with the contexts within which they operate.

We shall also see that the more one drills down into specific characteristics, the less we seem to have a good knowledge base, sound practices, and effective support and understanding. This is indicated by the opposing arrows of Figure 2.1.

General systems engineering is a long-established, well-regarded field with many researchers and practitioners having developed an abundance of standards, processes, and procedures. However, as we descend down the hierarchy to software and hardware systems engineering, we begin to hear complaints about whether or not the essence of systems engineering has been transferred to software systems engineering in particular. Quite a number of outspoken individuals decry the current state of the software systems engineering field as lacking the credibility, discipline, and effectiveness that would be expected of such a technical field.

Top class software development shops display high levels of sophistication, structure, and control throughout their software development life cycle processes, some accomplishing the highest levels of capability maturity. Perhaps the best-known example of a process improvement approach is the Capability Maturity Model Integration (CMMI) process introduced by the Software Engineering Institute of Carnegie Mellon University. A considerable amount of resources are available on their website [1]. It is interesting to note that, even though the Software Engineering Institute has developed and supported the CMMI process for over two decades, the approach is not limited to software development processes, but has expanded into such areas as system design, acquisition, security, risk, and process management. The International Systems Security Engineering Association (ISSEA) [2], in fact developed a Systems Security Engineering Capability Maturity Model (SSE-CMM–ISO/IEC 21827) [3]. However, judging from the ISSEA website, the organization appears to no longer be active. This is perhaps indicative of the low level of interest and support for the development of secure software systems.

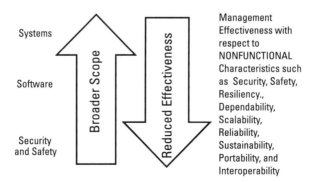

Figure 2.1 The relationship of management effectiveness to scope.

Yet, while those processes may meet the high capability standards that certification requires, the quality of software produced can still be deficient. For example, the software might not function properly (i.e., it does not meet all stated functional requirements) and may not satisfy nonfunctional requirements for security, resiliency, scalability, and so on.

What happens if we drill down one more level to the security and safety aspects of software systems? Safety-critical software systems are generally more solidly built; that is, they are more reliable and resilient than security-critical software systems. This is understandable because human life is often at stake. However, even safety-critical systems do not always achieve the highest standards, particularly with respect to their security attributes. Quite a large body of work has been developed over the past 15 years or so relating to building security into software, yet the acceptance of security requirements as part and parcel of a software system is low in comparison to the acceptance of safety requirements. That is not to say that there aren't any forward-thinking organizations that excel with respect to their security practices. There are. Some of them have had their achievements broadly publicized, such as in the Building Security In Maturity Model (BSIMM) report, which documents the security posture of firms that participated in the survey based upon a series of attributes [4].

As illustrated in Figure 2.1, we see that in general, certain aspects of the engineering of secure and safe software systems such as interoperability and portability diminish in the amount of attention paid to them as we enter into these areas of more limited scope.

What we are seeing is essentially a top-down process, with the message becoming more garbled and less intelligible as we approach more specialized and detailed areas. From reading a broad range of publications and attending a number of conferences and seminars on systems engineering, software engineering, and related fields, we see that, in general, systems engineers pay relatively little attention to software systems engineering; software engineers appear to ignore both the guidance that can be gleaned from systems engineering on the one hand, and the nonfunctional aspects of software systems, such as security, safety, and resiliency on the other hand. With such a top-down hierarchy, there is little chance of significant improvement in the overall security, safety, and resiliency of software systems. Indeed, it appears that we might be heading in the opposite direction.

Despite some excellent software engineering programs being offered by major universities, such as Carnegie Mellon University's Software Engineering Institute and the Stevens Institute of Technology, the field is still grossly underserved. For example, a Master of Software Assurance Curriculum [5] illustrates some significant advances that are being made in the academic arena. Unfortunately, the size and scope of these academic initiatives only begins to address the huge deficiencies in knowledge and training in this field.

The answer to the question of what needs to be done is fairly obvious: we need to ensure that the supporters of nonfunctional capabilities, such as security professionals, are involved at the beginning of the decision process and throughout the development life cycle of software-intensive systems. The challenge is not to determine what should be done—that has already been established—but how to get it done. With all the forces of time to market, cost savings, competitive advantage, budget constraints, complex functionality, and the like, working against the goal of safer and more secure software systems, it is unlikely that needed changes will happen of their own accord. There need to be incentives for those making such a radical change in approach, or disincentives for those not adhering to a mandate to change the approach to engineering safe and secure software-intensive systems.

Deficient Definitions

A significant detraction to solving the problem of attaining the needed level of proficiency in systems engineering, particularly with regard to software systems engineering, lies in the ambiguity of language used and the relative looseness of definitions of terms like systems engineering and software engineering, security engineering, and safety engineering. In [6], Allman describes, with considerable insight, the reasons for such ambiguity, as follows:

> Our normal human language is often ambiguous; in real life we handle these ambiguities without difficulty ... but in the technical world they can cause problems. Extremely precise language, however, is so unnatural to us that it can be hard to appreciate the subtleties. Standards often use formal grammar, mathematical equations, and finite-state machines in order to convey precise information concisely ... but these do not stand on their own ...

In its definitive guide, the International Council on Systems Engineering (INCOSE) [7] expresses this same concept with a specific focus on systems engineering (SE), as follows:

> One of the Systems Engineer's first jobs on a project is to establish nomenclature and terminology that support clear, unambiguous communication and definition of the system and its functions, elements, operations, and associated processes ... It is essential to the advancement of the field of SE that common definitions and understandings be established regarding general methods and terminology that in turn support common processes. As more Systems Engineers accept and use a common terminology, we will experience improvements in communications, understanding, and ultimately, productivity.

The INCOSE guide [7] then provides a list of definitions for frequently used terms, some of which will be repeated below.

A major contributing factor to deficiencies in software systems engineering appears to be the unwillingness or inability of specialists in one specific area (such as system engineers, software engineers, and hardware engineers) to communicate adequately. There is a particular need to improve communications between security and safety software systems engineers, as discussed in [8].

This lack of communication causes subject-matter experts and decision-makers in each narrow area of specialization to be at serious risk of missing important information regarding system requirements and other factors from those with the necessary expertise. This is particularly apparent during the critical requirements, validation and verification stages of the software, and hardware development life cycles. Such omissions can—and have—led to many inferior decisions.

Systems engineers dealing with entire systems seem to pay little attention to software considerations, as shown by the short shrift treatment of software topics in many systems engineering books, publications, and presentations. Conversely, with their concentration on the development, operation, and maintenance of applications and system software, software engineers tend to pay insufficient attention to many of the broader systems considerations, including platforms and infrastructures upon which applications and systems software operate as well as the human-system interactions. As a result, software products often do not exhibit the levels of confidentiality, integrity, availability, safety, interoperability, portability, scalability, resiliency, and recovery that should be incorporated into any critical system. This is mainly due to these nonfunctional characteristics of applications, systems software, firmware, and hardware not having been adequately accounted for in the design and development, and particularly, the integration of these systems.

Rationale

Clearly, the way in which one refers to various topics, subjects, and items can greatly affect how one deals with them. Therefore, we will now examine some definitions of systems engineering, as well as its components and subcomponents. We will look into definitions of applications, system software, software systems, and software engineering (or, more precisely in the last case, software systems engineering). It will be shown why the broad range of common usage of these definitions produces large gaps in understanding between what exists and what is needed.

We will then go through a similar exercise for security and safety engineering: first defining our terms and then describing why the philosophies and means of addressing issues differ so much between the two disciplines of software security engineering and software safety engineering.

We will investigate these gaps further and suggest some specific approaches and activities to address and close them. We will provide guidance for adhering to more structured systems-engineering processes for designing, building, testing, deploying, operating, and decommissioning safe and secure software-intensive systems. Furthermore, we will suggest how the numerous separate and independent silos can be brought together into a more collaborative environment so that each party can learn from the others and consequently arrive at much more productive and effective modes of operation and more acceptable results, as described in [8].

What Are Systems?

In order to come up with a viable definition for systems engineering, we must first develop and describe a structure that relates systems engineering to all the various components and subcomponents that arguably fall within its scope. We will select components that have particular relevance to this book from those shown in Figure 2.2. However, we need to always keep the remaining compo-

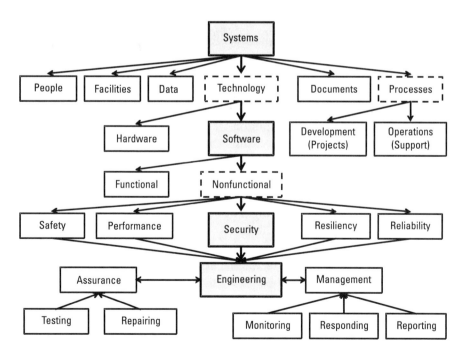

Figure 2.2 Structure and hierarchies of systems engineering.

nents and subcomponents in mind because they provide appropriate context for further discussions.

The hierarchy and relationships of components and subcomponents (as shown in Figure 2.2) provide a host of terms, and consequent definitions. At the top of the diagram, there is a box labeled "systems," and near the bottom is another box labeled "engineering."

First, let us examine the definitions for a *system* and for *engineering* separately, and then combined.

The IEEE Standard 610.12-1990, which has been replaced with IEEE/ISO/IEC 24765-2010 [9], defines a system as:

> ... a collection of components organized to accomplish a specific function or set of functions.

Christensen and Thayer [10] rework the above definition somewhat to come up with the following:

> A system is a collection of related or associated entities that together accomplish ... one or more specific objectives.

In the U.S. Department of Defense (DoD) text [11], the definition becomes:

> A system is an integrated composite of people, products, and processes that provide a capability to satisfy a stated need or objective.

When each of the above definitions are compared, the main difference is whether the goal is to satisfy a single need, function, or objective, or more than one of them. The IEEE definition takes care of both. However, the DoD definition also has merit because it lists the components as people, products and processes.

In Figure 2.2, the second row of boxes represents six elements that can make up a system, the first three of which correspond to the DoD definition. The elements include the following areas: people; technology (or products)—such as hardware and software; processes—development, operations; facilities; data; and documents.

As we proceed down Figure 2.2, the next two rows itemize the characteristics of the elements. For clarity, we only show those items that are of specific interest. However, it should be noted that each of the elements can be broken down, and that the breakdowns are also relevant to other areas of interest; this is not shown here in order to simplify the presentation. As a simple example, the category "data" can be broken down into structured and unstructured data. Our particular focus is on technology systems, and within that category we are focusing on software, although we shall touch upon other elements such as people (or human factors) and processes such as development and operations.

Software is further split into its functional and nonfunctional characteristics. This is a common way to divide software characteristics. However, as we will show subsequently, such a characterization has led to major gaps in how software systems are tested. In any event, the usual set of nonfunctional software characteristics is as follows:

- Security;
- Safety;
- Performance;
- Reliability;
- Compliance.

As the title of this book suggests, our main focus will be on the first two categories, namely, the safety and security of software systems.

The distinction between *safety-critical* and *security-critical* software systems will be examined in greater detail in Chapter 4. At this point, we present the definitions introduced by Boehm [12], as follows:

- *Safety-critical software:* the software must not harm the world.
- *Security-critical software:* the world must not harm the software.

That is to say, software systems that are judged to be safety-critical must not present a hazard to human life or the environment. This is an outward-looking perspective.

On the other hand, security-critical software systems are inward-looking to the extent that it is necessary to protect the software and any sensitive data such as nonpublic personal information and intellectual property from misuse, damage, or destruction from attacks by external or internal persons or computer applications.

As we shall see throughout this book, it is this divergence of viewpoint between software safety engineers and software security engineers that leads to major software systems engineering issues, particularly when there is a requirement to combine both safety-critical and security-critical software-intensive systems into what are increasingly termed *cyber-physical systems*, particularly by United States government agencies.

Near to the bottom of Figure 2.2, we show a box for "engineering." Engineering can be defined as follows:

... the creative application of scientific principles to design or develop structures, machines, apparatus, or manufacturing processes, or works utilizing them singly or in combination; or to construct or operate the same with full cognizance of their design; or to forecast their behavior under specific operating conditions; all as respects an intended function, economics of operation and safety to life and property [13].

At the bottom of Figure 2.2, we include activities relating to engineering, namely *assurance* and *management*. Assurance comprises testing and repairing. Management activities related to engineering include monitoring, analyzing, reporting, and responding.

As the reader can see, some of the boxes in Figure 2.2 are made up of solid lines, whereas other boxes are bounded by dashed lines. We will bypass the dash-framed boxes when using the hierarchy to designate a particular combination. Thus, we arrive at the term *software security engineering* rather than *technology software nonfunctional security engineering.*

Deconstructing Systems Engineering

The field of engineering has clearly had, and continues to enjoy, a very long and auspicious history that goes back over millennia. There is no question that the building of the Egyptian pyramids, the English Stonehenge, the Easter Island statues, the Mayan pyramids at Chichen Itza, Mexico, and other marvelous ancient constructions were amazing feats of engineering. Did those projects follow what we now would consider to be a systems engineering approach? The builders of these monuments clearly must have adopted some organizational structure and management processes in order to accomplish what they did. Archeologists have discovered highly sophisticated mathematical, architectural, and construction methods in documents used by engineers of these ancient monuments. However, the processes would not have followed the specific precepts of what we now call systems engineering in as formal a manner, although many of the components must have been utilized in order to accomplish these great works.

In modern times, the specific terminology for the discipline of systems engineering has only been in common use for a relatively short 70 years or so, although various documents attribute the creation of the term to different parties and at different times. Christensen and Thayer [10] quote Alberts [14], who asserted that the term was originated by a certain H. C. Hitch at Penn State University in 1956 [15].

The INCOSE website claims that the term *systems engineering* was coined by researchers at Bell Telephone Laboratories in the 1940s [16]. The INCOSE document [7] provides a timeline going back to the development of the Rocket locomotive by Robert Stephenson and Company in the 1820s, and includes a British team that analyzed the United Kingdom's air defense system in 1937. The INCOSE timeline also includes Bell Labs during 1939–1945, which agrees with the above claim. INCOSE attributes the invention of systems analysis to the RAND Corporation in 1956, which appears to differ from the Christensen and Thayer claim (with respect to who originated which terms). However, both agree that the term was originated in 1956. It is sufficiently accurate for our purposes—given the discrepancies in the literature—to state that systems engineering and systems analysis as we know now them originated in the mid-twentieth century and evolved through the latter half of that century.

A somewhat broad consensus has it that the discipline of systems engineering was driven in the 1940s and 1950s by the need to manage large, complex projects involving systems of systems, in which the properties of the sum of the parts of a system were often greater than the sum of the properties of the individual parts [17]. Much of the motivation for developing the art of systems engineering came from the requirements of the U.S. Department of Defense [11] and the National Aeronautics and Space Administration (NASA) [9]. We will, therefore, lean heavily on the works of these and other U.S. government agencies, as well as the publications of professional associations, such as the IEEE (Institute of Electrical and Electronics Engineers) [9] and INCOSE (International Council on Systems Engineering) [7], in our discussion of the attributes of systems engineering.

Today, there are many researchers and practitioners who are proud to number themselves among the ranks of systems engineers. Professional organizations, such as INCOSE and IEEE, which are a highly regarded and which produce impressive lists of publications and other valuable resources, have sprung up and prospered in response to the evolution of the field.

However, as discussed above, there appears to be a substantial gap between how software engineers view themselves with respect to systems engineering and how software systems engineers should integrate systems engineering into their own practices more extensively. This will be discussed in the following chapters. It is not that software developers do not follow structured practices equivalent to systems engineering guidelines—for the most part they do—particularly in the U.S. Department of Defense, where precise and detailed software development practices have evolved, as described in [11]. It is more that adoption of the formal processes of systems engineering is not generally accepted in what are considered related, but distinct, fields relating to software, safety, and security. For example, the system development life cycle (SDLC) has indeed been enthusiastically adopted by many modern software development shops, but is not usually embraced by smaller, less mature organizations because doing so introduces considerable overhead.

It is quite confusing to see so many combinations of the words *systems, software,* and *engineering.* What are the precise differences between systems engineering, software engineering, software systems engineering, and systems software engineering? Some of the differences may appear to be obvious, but others are not so obvious. When systems are aggregated into *systems of systems,* what terms should be used to describe the characteristics of the much more complex combinations of systems, which frequently exhibit behaviors beyond those of their individual component systems? The combination of safety-critical systems and security-critical systems, such as is occurring with the so-called smart grid, produces issues well beyond the reach of the individual systems.

In order to get a handle on these issues, we use the elements of the table in Figure 2.3 to assist with explaining the definitions and the relationships among components. For example, we see "systems engineering" in the first line of the table, "software engineering" in the second line, and "software systems engineering" in the third line. If we were to reverse Columns 2 and 4 in the third line, we would get "systems software engineering," which has a completely different meaning from "software systems engineering." Systems software engineering relates to the engineering of systems software. Software systems engineering refers to the application of systems engineering practices to all types of software-intensive products. Systems software is a category of software that lies below the applications software layer and performs many of the administrative functions

Secure/Safe	Software	Security/Safety	Systems	Engineering
			Systems	Engineering
	Software			Engineering
	Software		Systems	Engineering
		Security/Safety		Engineering
Secure/Safe			Systems	Engineering
	Software	Security/Safety	Systems	Engineering
Secure/Safe	Software			Engineering
Secure/Safe	Software		Systems	Engineering
	Software	Security/Safety		Engineering
	Software	Security/Safety	Systems	Engineering
Secure/Safe	Software	Security/Safety	Systems	Engineering

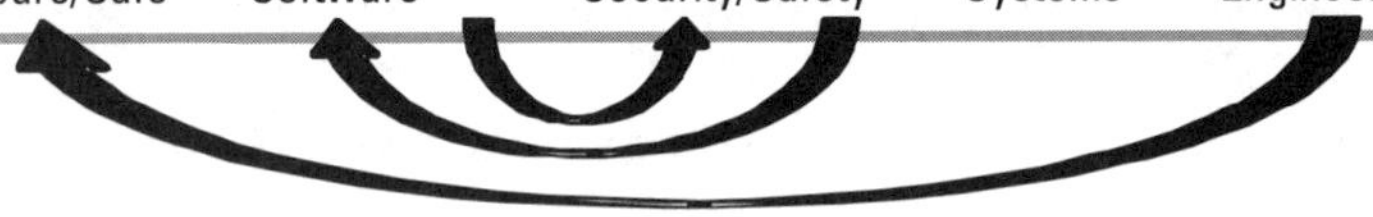

Figure 2.3 Terminology and combinations of terms.

that support the applications. It includes operating systems and utilities, such as sort algorithms.

Some might consider such differentiations as those mentioned above and illustrated in Figure 2.3 to be splitting hairs, but terminology that is confusing and ambiguous will almost invariably lead to misunderstandings or worse, as was suggested above. Such a situation arises when researchers and practitioners miss whole bodies of applicable work because calling the same item by partially or completely different names results in their searches for relevant information being incomplete or, worse yet, inaccurate and misleading. We have already discussed how software engineers often do not see themselves as systems engineers and vice versa, but this is merely on the surface. As we dig deeper, we will find that such high-level perceptions lead to problems with the omission of important requirements for complex systems that are commonly developed and implemented today.

What Is Systems Engineering?

At this point, we will tackle the seemingly simple task of defining systems engineering. There are many definitions of this term, each of which differs somewhat from the other, sometimes minimally; in other cases, the differences are substantial.

One widely-quoted definition of systems engineering is the one in the DoD text [11], namely:

> … an interdisciplinary engineering management process that evolves and verifies an integrated, life-cycle balanced set of systems that satisfy customer needs.

An important aspect of this definition is that systems engineering is a process involving a variety of professional disciplines. The sole objective of this process is to create systems that meet users' needs. The author does not argue with this definition, but will raise questions with respect to the real-world implementation of the process. For example, are all necessary disciplines included at appropriate stages of the process? It will be asserted that they are not. Have the needs of customers or nonhuman users been adequately identified? It will be claimed that users likely only express the functional requirements of which they are personally aware or that directly affect them, and that they omit a host of requirements that they either don't know about or don't care about.

Another definition of systems engineering is:

... an interdisciplinary field of engineering that focuses on how complex engineering projects should be designed and managed over the life cycle of the project. [19]

And yet another definition [10, p. 8] is:

... the practical application of the scientific, engineering, and management skills required to transform a user's need into a description of a system configuration that best satisfies the need in an effective and efficient way.

Clearly these definitions suggest a life cycle process that begins with the identification of user needs and should end up with one or more systems that meet those needs. The success of systems engineering in general depends on efforts to organize system design and development processes that function effectively within and between organizational units because the right level of management and control functions exists. In the following section, we examine the various stages in the process and describe some effective controls for the systems creation environment.

Systems Engineering and the Systems Engineering Management Process

As mentioned above, the pioneers of systems engineering were involved in large government (or public sector) programs relating to space exploration and military systems. We will, therefore, place emphasis on those sources initially, and then look at how the systems engineering (SE) approach has been adopted by others, particularly in the nongovernment or private sector.

The definitions of public and private sectors can vary with country. In the United States, "public" is generally equivalent to "government" and "private" means "nongovernment," although nonprofit, academic, and religious organizations are often recognized to be different. In Great Britain and many other European countries, a third sector is defined. This refers to nongovernment, nonprofit organizations. In other countries, such as Japan and India, they define a "joint sector," which refers to industries or companies owned and/or run by both the public and private sectors [20]. For the purposes of this book, we differentiate between government and nongovernment sectors.

Christensen and Thayer [10] list the following systems engineering functions:

- *Problem definition* is the determination of customer requirements with respect to product expectations, needs, and constraints.

- *Solution analysis* is the analysis of options satisfying requirements and constraints and selection of the optimal solution.

- *Process planning* is the prioritization of major technical tasks and the effort required for each task and the determination of potential project risks

- *Process control* is the establishment of control methods for technical activities including the measurement of progress, review of intermediate products, and initiation of corrective action when necessary.

- *Product evaluation* is the determination of quality and quantity through testing, demonstration, analysis, examination, and inspection.

These life cycle functions are common throughout software systems engineering projects and their subcomponents. The latter deal with such attributes as security and performance. It is, however, quite confusing because different types of projects result in varying emphasis on specific activity phases and the terms used, particularly when comparing DoD terminology with others, which often results in even more confusion.

In simple terms, requirements are obtained and converted into some form of system or other by means of a series of activities making up a project. Due diligence suggests that, prior to release, a system needs to be thoroughly tested to ensure that it meets original requirements. This sounds simple, but, as we shall see, if it were that simple then there would be universal standards, which, if adhered to, would presumably result in systems that meet their requirements perfectly every time. Clearly, we do not see such a result and this is because there are always tradeoffs that have been made due to constraints on resources and funding, time until delivery, and the ability to do as good a job as possible. A project is always a tradeoff among timeliness, quality, and cost; you can try to optimize two out of three, but it is not feasible to optimize all three for the same project. With respect to the *project management triangle*, practitioners will summarize this as: fast, good, cheap—pick any two [21]! A major exception to this restriction appears to be well-funded government contracts, where cost overruns and time delays are common and seemingly acceptable, although much of the reason for this may lie in the governance model in which politicians with little technical knowledge—but a clear political agenda—have responsibility for oversight of the management of the project.

It is ironic that the origins of systems engineering and by far its most rigorous methodologies have emanated from military and space programs, both of which are government sponsored, and yet these projects appear to frequently suffer huge overruns in time and money and often lack critical features that were specified in the requirements documentation. The same appears to be quite common in academia, where the product itself is often subservient to the potential accolades and kudos to be gained from well-accepted research papers and publications in quality journals. That is not to say that the private sector is doing any better a job in general, but where there is effective management and governance, large complex projects do get done on time, fall within budget, and exhibit high quality.

Nevertheless, the private sector can learn a great deal from the methodologies created by research organizations for government, as long as effective governance and strong project management are implemented. We will, therefore, proceed with further descriptions of systems engineering management derived from publications by engineering groups, such as the Institute of Electrical and Electronic Engineers (IEEE) Computer Society, and the U.S. Department of Defense (DoD) initiatives.

The DoD Text

Perhaps one of the most detailed and complete descriptions of the systems engineering process is that utilized by the DoD. The document, *Systems Engineering Fundamentals*, was prepared and published in 2001 by the DoD Systems Management College [6], and is available online for everyone to read.

For those, such as this author, whose information technology and information security career have been in the private sector, the terminology and emphasis of the DoD approach can appear quite foreign. The management process is seen as being highly regimented, with little opportunity for variations. This likely stifles innovation, as the quotation by Steve Jobs at the beginning of this chapter would suggest. Of course, given the types of systems developed for the military, perhaps the last thing that one wants to see is unbridled creativity, which is nonetheless often the hallmark of leading-edge commercial systems ventures.

As will be discussed later in this book, this dichotomy of approaches between military and commercial systems engineering carries through to many other aspects of the systems. In particular, military systems have (up until recently) emphasized safety, reliability, and predictable operation above all else, whereas commercial systems generally look to beef up security (although vehicle control systems do focus on safety). Going forward, however, the importance of security, especially for network-centric systems, is being recognized and accounted for in all areas of endeavor.

Another Observation

In researching for this book, the author became fascinated with the singular lack of reference to the DoD standards and practices in books and articles about nongovernment, especially nonmilitary, systems and software engineering. It is as if a whole body of knowledge relating to military and other government systems and software engineering has been excluded from consideration in those publications aimed at a general audience. This is a shame, to say the least, as such knowledge transfer would help all involved parties.

More on Systems Engineering

Remaining on this topic of the differences in approaches between various public-sector and private-sector entities, we will briefly look at how the emphasis of systems engineering management approaches differs between groups and how the difference in emphasis is at once considered appropriate for the types of system that have traditionally been within their purview, but, on the other hand, hampers each group from benefiting from the knowledge and skills of other groups. Some professional organizations, such as INCOSE, make creditable attempts to be all-inclusive and try to encourage broader participation, yet many other groups remain quite insular.

It is suggested, therefore, that the reader delve into a broader range of bodies of knowledge to gain from the wisdom developed within the various silos, which unfortunately do not communicate adequately. Some of the differences that you are likely to discover have to do with the more highly developed governance models that government entities have created and their somewhat obsessive concentration on requirements going in and on verification and validation coming out of the process. Not to take anything away from focusing on these important areas, but much of the private sector seems unable to afford the luxury of such time and money consuming efforts. Despite this drawback, it should serve us well to take a brief tour of the DoD approach and compare and contrast various components.

The Systems Engineering Process (SEP)

The DoD document [11] describes the SEP as "... a comprehensive, iterative and recursive problem-solving process, applied sequentially top-down by integrated teams." A significant point here is the use of the term "top-down," clearly pointing to the existence of a well-defined hierarchical structure. While the corporate world also has, for the most part, clearly laid-out organizational structures, there is generally more room for collaborative teamwork within the corporate structure. In fact, in many highly creative systems and software development shops, particularly with start-ups and fast-moving companies, designers and developers are given considerable freedom and latitude when it comes to problem-solving and putting together systems requirements. Of course, this latter environment has both positive and negative characteristics, and these would not likely work for weapons systems, for example. Larger, more mature private institutions typically have many formal processes and management controls in place, and may be more readily compared with the military, which starts out being large and formal.

When it comes to developing system requirements, government—particularly the military—have well-defined user populations that may have little direct input into defining initial requirements, although they will probably be involved in testing systems in the final stages and providing feedback for tweaking the systems. While the DoD document talks about customers, these recipients of the end product do not necessarily have much choice about whether or not they are comfortable with using the resulting systems, despite any deficiencies in their design or manufacture.

That is to say, the usual competitive marketplace—while existing among bidders for the military contracts—is not necessarily in force with respect to end users when it comes to choosing system features and functionality, for example.

The difference between customers who have free choice and those who don't affects their influence on the design of a system and its use. Here is a real-world example. The information technology department of a highly-structured financial institution (for which both Asian and American traders worked) designed and developed trading systems based on some minimal set of requirements obtained from some of the traders. After the trading system has been installed, the financial institution insisted that traders use the system as delivered, whether or not it fully met the traders' needs. When the same system was presented to traders in the United States (who typically consider themselves independent entrepreneurs as opposed to subservient employees), they refused to use the system. It was very apparent that cultural differences between traders in each country had a lot to do with how systems were designed and developed and whether or not they were accepted. To some extent, the same observation holds true with government agencies and internal corporate systems. However, when it comes to commercial systems, the marketplace determines what gets used and what languishes.

The formal requirements process in the DoD SEF document [11] indicates that the system operator is the key customer, and that customers are required to provide the basic needs for the system. According to the DoD SEF [11], operational requirements should answer the following questions, some of which have been slightly modified from the original:

- Where will the system be used?
- How will the system accomplish the mission objective?
- What are the critical system parameters to accomplish the mission?

- How should the various system components be used?
- How effective and efficient must the system be in performing its mission?
- How long will the system be in use?
- In which environments will the system be expected to operate effectively?

The DoD SEF [11] lists the following system requirements:

- *Customer requirements* are the expectations of systems in terms of mission objectives, environment, constraints, and measures of effectiveness and suitability.
- *Functional requirements* are the necessary tasks, actions, or activities that must be completed.
- *Performance requirements* are the extent to which mission or function must be executed with respect to quantity, quality, coverage, timeliness, and readiness.
- *Design requirements* are the "build to" requirements for products (e.g., hardware, software) and "how to execute" requirements for processes.
- *Derived requirements* are the implied or transformed from higher-level requirements.
- *Allocated requirements* are established from division of high-level requirements into a number of lower-level requirements.

It is noteworthy that the above requirements list makes no explicit mention of either security or safety. These latter factors—along with others relating to such characteristics as resiliency and interoperability—would likely be subsumed within the performance requirements, were they to be considered at all. Nevertheless, the lack of appearance of these nonfunctional requirements is not surprising because broader publications in systems engineering also tend to omit them. It is usually only when, for example, there are special issues of magazines and journals related to such fields are systems engineering, electrical engineering, or other branches of engineering focused on software, security, and safety (as well as books that specifically cover software, security and safety engineering, and related subjects), that the reader of the broader engineering publications gets to learn more about these topics.

Summary and Conclusions

In this chapter, we have addressed some of the confusion in regard to the definitions of systems and systems engineering, and we have discussed some of the issues that arise in the systems engineering process. The reader can access some excellent guides and handbooks on systems engineering, some of which (such as the DoD and NASA documents referenced in this chapter) are available in the public domain at no charge, to learn more about specific aspects of the subject.

The main take-away from this chapter should be recognition that there is inadequate sharing among various government and nongovernment players and among different subcategories, such as those relating to security and safety.

A goal of subsequent chapters is to encourage the engagement of diverse groups, who today do not communicate sufficiently, and also to achieve some measure of consistency across fields.

Endnotes

[1]   Available at http://www.sei.cmu.edu/cmmi/, last accessed on July 8, 2012.

[2]   Available at http://www.issea.org, last accessed on July 8, 2012.

[3]   Available at http://www.sse-cmm.org/issea/issea.asp Accessed on July 8, 2012.

[4]   Available at http://bsimm.com, last accessed on July 8, 2012.

[5]   Available at http://www.cert.org/mswa/, accessed on July 8, 2012.

[6]   Allman, E., "The Robustness Principle Reconsidered," *Communications of the ACM*, Vol. 54, No. 8, 2011, pp. 40–45.

[7]   Haskins, C. (ed.), *Systems Engineering Handbook: A Guide for System Life Cycle Processes and Activities*, San Diego, CA: International Council on Systems Engineering (INCOSE), 2011.

[8]   Axelrod, C. W., "Applying Lessons from Safety-Critical Systems to Security-Critical Software," *2011 IEEE LISAT (Long Island Systems, Applications and Technology) Conference*, Farmingdale, NY, May 2011, published on the *IEEE Xplore* website.

[9]   IEEE/ISO/IEC Standard 24765:2010, *Systems and Software Engineering—Vocabulary*, 2010. Available at http://ieeexplore.ieee.org/xpl/mostRecentIssue.jsp?punumber=5733833, accessed on July 8, 2012 (registration required to obtain document).

[10] Christensen, M. J., and R. H. Thayer, *The Project Manager's Guide to Software Engineering's Best Practices*, Los Alamitos, CA: IEEE Computer Society, 2002.

[11] United States Department of Defense (DoD), *Systems Engineering Fundamentals*, Fort Belvoir, VA: Defense Acquisition University Press, 2001. Available at http://www.dau.mil/pubs/pdf/SEFGuide%2001-01.pdf, accessed on July 8, 2012.

[12] Boehm, B. W., *Characteristics of Software Quality*, New York: North-Holland Publishing Company, 1978.

[13]   This definition of *engineering* is available at http://en.wikipedia.org/wiki/Engineering, last accessed on July 8, 2012. The definition is attributed to the American Engineers' Council of Professional Development (ECPD), which is the predecessor of the Accreditation Board for Engineering and Technology (ABET).

[14]   Alberts, H. C., "System Engineering—Managing the Gestalt," *System Engineering Course Syllabus*, Fort Belvoir, VA: Department of Defense Systems Management College, 1988.

[15]   The specific reference to H. C. Alberts can be found in Chapter 1 of *The Project Manager's Guide to Software Engineering's Best Practices* [5]. This chapter is posted in full on the Wiley website at http://media.wiley.com/product_data/excerpt/96/07695119/0769511996.pdf, last accessed on July 8, 2012.

[16]   "A Brief History of Systems Engineering," is available at www.incose.org/mediarelations/briefhistory.aspx, last accessed on July 8, 2012.

[17]   A *system of systems* is defined in the INCOSE document [2] as applying "to a system-of-interest whose system elements are themselves systems; typically these entail large scale inter-disciplinary problems with multiple, heterogeneous, distributed systems."

[18]   National Aeronautics and Space Administration (NASA), *NASA Systems Engineering Handbook, NASA/SP-2007-6105 Rev 1*, Washington, DC: NASA, 2007. Available at http://www.tsgc.utexas.edu/challenge/PDF/NASA-SystemsEngrHandbook.pdf, accessed July 8, 2012.

[19]   This definition of *systems engineering* is available at http://en.wikipedia.org/wiki/Systems_engineering Accessed on July 8, 2012.

[20]   Available at http://en.wikipedia.org/wiki/Voluntary_sector, last accessed on July 8, 2012.

[21]   For an overview of the project triangle, see http://en.wikipedia.org/wiki/Project_triangle, last accessed on July 8, 2012.

3

Engineering Software Systems

Software engineering is still an oxymoron [1]
—John Pescatore, Gartner Inc.

... software engineering, as originally envisaged, does not yet exist [2]
—David Lorge Parnas, Middle Road Software

Introduction

Software is different from most other products and services acquired by consumers, customers, and organizations. It is arguably more like a fuel than it is a product because all systems that include software as a component require an infusion of software in order to operate. Yet, unlike a fuel, software doesn't get depleted with use. Application software also differs from a fuel in that it is the product that you, the so-called user, usually interact with, rather than something that is somehow activated in the background, as is a fuel. However, as with a fuel, software is not self-sufficient. It needs a context within which to perform. Out of context, it is useless. If you obtain a disk containing software or you download software over a communications link, the software has to be used on a system comprising particular equipment (or hardware), such as computer processors, storage devices, and communications networks, and (for applications) system software and utilities, all of which support software products but themselves do not have any end-user functions without the requisite application software.

Other characteristics of software products and the data that they handle have a profound impact on security, safety, resiliency, and similar nonfunctional software system attributes. One is that software can be duplicated for extremely low incremental cost and can be modified or exploited at will (although

29

doing so will usually negate any license terms and warranties), often without the changes or attacks being detected and without the perpetrators having to be in the same physical location as the original software, as long as they have a means of connecting to or accessing the software and data. Another characteristic of software is that customers generally license rather than own commercial-off-the-shelf (COTS) software (as opposed to home-grown software, which is owned by the manufacturer or its designees), so that one is limited as to what one can do with COTS software in terms of alteration of the program code and reselling the resultant product. One exception to this rule is open-source software, which is owned by no one and maintained by a community. However, open-source license agreements have their own idiosyncrasies that must be considered, as in [3].

It might seem reasonable, therefore, to explain the reticence of some systems engineers, comfortable in fields such as mechanical, civil and electrical engineering, and architecture, to becoming involved with and dealing with software and its related issues due to the unique characteristics of software as described above. As mentioned in [4], such traditionalists are likely to be more comfortable in the physical world (as opposed to the virtual world), even though they will likely use personal computers for various purposes such as emailing, social networking, spreadsheet calculations, and word processing. However, this latter involvement is not the same as that of incorporating industrial-strength software into the actual systems that they are engineering. By giving short shrift to software and its implications, such systems engineers will likely omit serious software considerations when designing and building systems and when interfacing their traditional physical systems with the increasingly pervasive virtual world.

On the other hand, today's so-called software engineers often do not have a formal grounding in software engineering but enter the software systems engineering field directly through basic training, in college or commercial training programs in software programming skills, and on-the-job experience in a development shop, with occasional certifications in specific vendors' products for the use of specific platforms and infrastructures.

That is not to say that formal project management methods are not fairly common in software development projects. They are particularly prevalent in large software manufacturing entities. In this chapter, we present a number of well-formulated examples of such approaches that are described in the literature. In reality, the methods and practices of systems engineering project management have existed in the software development space for decades. However, endorsing and practicing project management alone is not the same as following the philosophy and practices of the systems engineering approach. The structure and rigor to be found in systems engineering is more prevalent in the development of software for military systems, avionics, space programs, and the like, than it is with noncritical commercial systems. However, certain critical sectors, such as financial services, do use formal techniques to ensure the correct functionality of their operational computer-based systems through which trillions of dollars are processed every day. (✳)

The Great Debate

There are a number of individuals, such as John Pescatore and David Lorge Parnas (quoted above), who believe that the field of software systems engineering has generally failed to absorb the rigorous practices and disciplined processes and procedures of systems engineering. This view is reinforced by frequent reports in the press of successful attacks against software systems and the seeming inability of many defenders to protect systems against such attacks effectively. It is not so much a matter of not trying to systematize software systems engineering. To the contrary, a huge amount of research and practical effort has been expended in this attempt. As will be shown in this book, it is in large part due to differences in knowledge, background, experience, capabilities, and goals of those individuals who are doing the work. Software remains as much art as science, and few practitioners developing today's huge, complex software systems have the personal breadth or the interest in reaching beyond the particular needs of their jobs. On the other hand, the majority of those with security, safety, privacy, dependability, performance, and resiliency backgrounds do not have a good grasp of applications software and the software development process; they often seem incapable of explaining their requirements to developers and especially to the users, managers, and customers for whom the software systems are meant.

Nevertheless, despite the apparent flagrant disregard among many creators of software systems of formal approaches of systems engineering and other engineering disciplines, there appears to be increasing recognition of the need to build into software systems such attributes as security (i.e., confidentiality, integrity, and availability), safety, survivability, interoperability, portability, sustainability, and resiliency. The rationale for this growing appreciation for nonfunctional requirements is readily apparent; the earlier in the development life cycle that such factors are considered (and errors or omissions found), the less expensive it will be to include their requirements. Yet attempts to ensure that the desired levels of these nonfunctional factors are incorporated into software systems seem not to have gained sufficient traction to overcome the recent rapid surge in reported attacks, particularly if one believes that there is a relationship between more formal methods and improvements in the quality of software, which is the fundamental assertion of this book. This growth in exploitable and exploited vulnerabilities continues despite the efforts of many researchers and practitioners to include security, safety, and resiliency into the software system development life cycle processes, suggesting that we need both new and improved approaches to engineering such systems.

The raging debate of the moment revolves around whether software engineering is, or ever will be, accepted as an engineering discipline. However, there remains a question as to whether or not it really matters whether one can pigeon-hole or label software engineering as engineering. We must ask what differences in approach are suggested by particular definitions and whether or not those differences are, in fact, material.

Some Observations

As in the previous chapter, we will use the hierarchy of terms presented in Figure 2.2 in order to assist in our defining software engineering, and to justify using the preferred term software systems engineering.

As we descend one level from systems to software (as well as hardware) systems engineering as depicted in Figure 2.2, we are confronted with complaints about whether or not the essence of systems engineering has indeed been handed down to software systems engineering. Many outspoken individuals decry the current state of the software engineering field as lacking the discipline and effectiveness that is common for systems engineering. Best-of-class software development shops display high levels of structure and control throughout their software systems development life cycle processes, some accomplishing very high levels of capability maturity. Yet, while the processes may meet high standards, the quality of the software that is developed may occasionally be deficient because the software often does not function properly (i.e., does not meet all stated requirements) and/or lacks the security, resiliency, scalability, and so on, that holistically designed software should have. A particular concern is that software designed for one platform and then transported to another platform may not be fully compatible with the second environment, due in part to the main testing effort having been focused on software running on the primary platform.

In general, certain attributes of the engineering of secure and safe software systems diminish as we drill down into areas of more limited scope. It is essentially a top-down process, with the message becoming more garbled and fuzzy as we descend into the more specialized areas. Thus, we observe that systems engineers may pay relatively little attention to software engineering. Software engineers, in turn, may well ignore (or more likely, pay minimal attention to) the so-called nonfunctional aspects of software systems. Ironically, the nonfunctional requirements are often of more significant concern to systems engineers than to software engineers, as the former will likely take a broader, more holistic view of the systems that the software is a part of. With a top-down hierarchy, there is very little chance of much improvement in the overall security, safety, resiliency, and performance of software systems. Indeed, it often appears that we are charging in the opposite direction until some limiting factor or costly incident makes the engineering community focus on a particular attribute that dramatically comes to the fore.

The answer to the question of what needs to be done is apparent—we need to have the guardians and sponsors of nonfunctional capabilities involved from the beginning of the decision process and active throughout the life cycle of designing, developing, implementing and operating software systems. Such advocates of nonfunctional soundness need to be fully and visibly supported and empowered by senior management in order for them to be effective in their missions. If management only pays lip service to aspects like security, then the appropriate level of attention will never be given to a particular attribute until its high-impact failure draws questions regarding management's effectiveness and their ability to control the environment. By then, of course, much of the damage will have already been done.

The challenge is not so much a matter of what should be done, but how to get it done. In reality, we are pressured by factors such as time-to-market, cost reductions, competitive advantage, budget constraints, complex functionality, and other compelling features, which work against goals of safer and more secure software systems. In such an environment, it is unlikely that needed changes and improvements will just happen. There need to be specific and direct incentives for making such major changes in approach and investment in nonfunctional attributes. Alternatively, there should be disincentives for those not adhering to management mandates to change the approach and include nonfunctional attributes into the software system development life cycle.

Rationale

As stated previously, one's approach to various subjects is often greatly affected by how one's terms are defined. Therefore, in this chapter, we examine definitions of software and software engineering—or, more precisely, software systems engineering. It will be shown that the broad range of usage of these definitions has itself produced large gaps between what is really needed to achieve an optimal balance among all contending attributes and what exists today.

Subsequently, we will go through a similar exercise as we did previously for other terms, but this time specifically for security engineering and safety engineering: first defining our terms, and then describing why the philosophies and means of tackling issues, which often differ so much between the two disciplines addressing security and safety, create so many problems.

We will further investigate these differences and suggest some specific approaches and activities that will help address the gaps and, it is hoped, narrow or close them. We will provide guidance for adhering to more structured systems-engineering processes in the design, building, testing, deploying, operating, and decommissioning of safe and secure software systems. Furthermore, we will suggest how currently-independent silos can be brought together in a more collaborative atmosphere so that each can learn from the other and consequently arrive at a much more productive and effective cooperative mode of operation. It is not sufficient to merely suggest what will be good for an organization, industry, country, or the global economy. Such normative appeals are seldom convincing, and few (if any) provide the practical guidance needed to make the desired actions happen. Here, we will demonstrate the benefits of including the nonfunctional attributes early in the life cycle using arguments that will appeal to nontechnical senior management, policy and law makers, auditors, and enforcers. What it comes down to is that suggestions offered in this book not only make technical and engineering sense, but also offer a sizable return on investment.

Understanding Software Systems Engineering

One reason why some might claim that software engineering is deficient is that the term itself is too general and not well understood by researchers and practitioners alike. One might readily presume that the term software engineering simply means applying engineering principals and methods to the software development life cycle (SDLC). However, software engineering can mean much more than merely the software manufacturing process. It also involves installing the software on a variety of platforms and infrastructures, and integrating software products into existing environments and into systems of systems.

If one accepts a simplistic definition of software engineering, one might reasonably presume that the inadequacies in the practice of software engineering have been fully addressed already by many who manufacture software for internal use or sale. However, all the issues clearly haven't been resolved, as witnessed by the increasing number and magnitude of computer system failures and the rapid escalation in the number and impact of successful compromises of today's systems.

Deconstructing Software Systems Engineering

We will now spend some time and space trying to tease out workable meanings from the multitude of misunderstandings and misinterpretations that haunt software engineering and its related areas.

In the first place, when we refer to software engineering, we likely mean the design, development, manufacture, and operation of products that are more extensive than mere standalone computer programs. We should use the term software systems engineering, because software programs, whether end-user applications, utilities, or systems software, do not and cannot operate in isolation. Software programs exist within broader systems and are frequently only able to run on specific platforms and within particular infrastructures. Software is only one component of an overall system comprising hardware, communications networks, people, and processes; but even the apparently straightforward term *software systems engineering* remains somewhat ambiguous. When using the term software systems engineering, do we mean the engineering of software systems or systems engineering methods applied to software development, or should we be even more explicit by expanding the expression to systems engineering methods applied to software-intensive systems?

While the reader might consider such a distinction to be overly specific and too exacting, it will be shown that many of today's software issues, as they relate to ensuring that software systems are safe, secure, resilient, interoperable, and the like, actually emanate from lack of rigor in defining terms. Defining all of the many characteristics of software is not an easy process, nor is going through the exercise guarantee that all outstanding issues will be resolved. However, we will demonstrate that some definitions in this area are often in conflict, and that such conflicts result in researchers and practitioners frequently missing the point. We will, therefore, try to shed some light on the murkiest areas by defining our terms as explicitly as possible.

What Is Software?

Before we discuss the meaning of software engineering, we need to work on defining "software." In simple terms, software consists of computer program code, usually written in high-level computer languages, such as C, C++, and Java. The resulting source code, of which much software is comprised, is then converted into code that is understood by computers by interpretation or compilation. Interpreters convert the source code into machine-readable code each time the software is run, whereas compilers produce machine-readable object code that is installed only once and is only recompiled if there is a change in the source code. Compiled programs tend to be more efficient, especially when there are few changes being made to the software relative to the frequency of operation. Where efficiency is paramount, developers might write code directly in machine language and optimize the software's run-time demands on system resources.

What Are Software Systems?

There are many flavors of software, such as end-user applications, systems software, embedded software, and firmware. Here again, definitions abound and their use is often quite sloppy, which leads to confusion, errors, and omissions. In Figure 3.1, we see a typical traditional software structure where applications run on systems software, which in turn, runs on hardware. The diagram shows layers of different types of software. In the minds of the general public, software is synonymous with end-user applications or "apps." Such software is often created by applications developers who may have minimal knowledge of underlying platforms and infrastructures, and may relate to the latter through standard application programming interfaces (APIs), which allow developers to create application functionality without having to worry about the systems' internals. This has the advantage of directing the development of highly technical systems software to those with particular expertise in such areas.

However, before going further, let us define software in its various forms and relate it to software architecture, as illustrated in Figure 3.2.

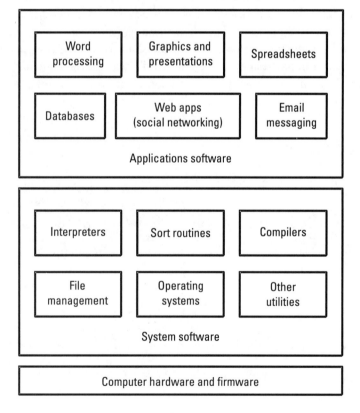

Figure 3.1   Typical IT system architecture.

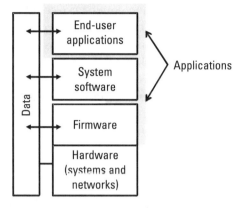

Figure 3.2 Typical applications software environment.

The terms software and computer program are generally considered to be synonymous, and most people do not agonize over the specific definitions used. However, as software is increasingly embedded into computerized equipment (or hardware), it is becoming much more difficult to differentiate among the various realizations of software. In some sense, it can be detrimental to one's understanding of systems if definitions are overly specific. This is especially true when we consider that software is essential for the creation of computer hardware consisting of transistors, other electronic components and the circuitry that are combined to form integrated circuits. Physical units are carved out of metals, plastic, ceramics, and increasingly sophisticated composite materials using computer-controlled software-driven machine tools. With the advent of three-dimensional printing, physical components are now being created (albeit mostly in laboratory environments) directly from software systems, jumping ahead technically beyond software-controlled machine tools used for carving out physical components, although limited in the types of component that can be created. Consequently, with software infiltrating into practically every aspect of our physical and informational lives, it is no longer suitable to distinguish software for independent consideration. Some software actually generates other software and is able to create physical objects (as with three-dimensional printing). Software is a rapidly growing part of an increasing number of systems used in our everyday lives.

It is interesting to note that references to software in software engineering books and articles are often cursory, with authors presuming that we all know what software is. For example, Braude and Bernstein [5] point out that "the goal of software engineering ... is the creation of software systems ..." without saying exactly what they mean by "software." Of course, one can usually infer the meaning that they assign to the term from its use throughout the book, but that is not the same as providing a specific definition. Similarly, according to Christensen and Thayer [6], "[t]he dominant technology in many modern products is software ... [which] often provides the cohesiveness, control, and functionality that enable products to deliver solutions to customers."

It might be true, at one level, that we have an intuitive sense of what software is. However, in this book we take a different viewpoint and go as far as to say that by omitting the definition of software, researchers are creating ambiguity and encouraging misinterpretation, as Allman points out in [7]. Software is complex and dynamic and—in order to deal with issues confronting the creation, implementation, operation, and disposal of safe and secure software systems—we need to come up with precise, yet somewhat flexible, definitions.

First, let us repeat some of the common definitions of applications and system(s) software. The following definitions, except as otherwise noted, are from the Open Projects website [8]. The definition of software is as follows:

> Software is a generic term for organized collections of computer data and instructions, often broken into two major categories: system software that provides the basic non-task-specific functions of the computer, and application software which is used by users to accomplish specific tasks.

The above definition raises an issue as to whether data should be incorporated into software or treated separately. It is held here that software and data are separate entities, which have different roles within software systems, even though one can hard code data into an application—a practice that is better avoided if possible because a change in data might then require software modification, unit and integration testing, and recompilation, which is time-consuming and somewhat risky. If an entry in a data table can be changed instead, it is often much simpler to implement, even though it may still be necessary to test the system with the change in data.

Software has both functional and nonfunctional attributes, as will be discussed when we examine security and safety engineering. Data, on the other hand, can relate to or be independent of software and often does not need modifications to software when the data changes, although reformatting the same data or converting the data to a different file or database system will undoubtedly require changes to software. In Figure 3.2, we specifically show data as being separate from software. We also illustrate that—while virtually all end-user software programs are generally considered to be applications—because they perform tasks as opposed to controlling hardware or equipment, systems software and firmware can also perform end-user tasks such as utility programs that sort data. Hence, some subset of systems software and firmware can be considered applications according to our definitions. This differentiation becomes particularly important when comparing and contrasting application-related and system-related software.

According to the Open Projects Web site [9], we have the following definition of systems software:

> System software is responsible for controlling, integrating, and managing the individual hardware components of a computer system so that other software and the users of the system see it as a functional unit without having to be concerned with the low-level details such as transferring data from memory to disk, or rendering text onto a display. Generally, system software consists of an operating system and some fundamental utilities such as disk formatters, file managers, display managers, text editors, user authentication (login) and management tools, and networking and device control software.

First we should note that, in some cases, writers refer to either system engineering or systems engineering. While systems engineering appears to be the preferred term, the other version, namely, system engineering, is used here whenever a quoted source uses that term. Also, we shall refer to authentication as a security service later in this book. While software realizations of authentication systems are common, authentication can include hardware that is independent of the hardware platform used by the applications. Biometrics readers and card scanners are examples of such equipment and are a good reason for precision in our definitions.

We have the following definition of application software per the Open Projects Web site [10]:

> Application software ... is used to accomplish specific tasks other than ... running the computer system. Application software may consist of a single program, such as an image viewer; a small collection of programs (often called a software package) that work closely together to accomplish a task, such as a spreadsheet or text processing system; a larger collection (often called a software suite) of related but independent programs and packages that have a common user interface or shared data format, such as Microsoft Office, which consists of closely integrated word processor, spreadsheet, database, etc.; or a software system, such as a database management system, which is a collection of fundamental programs that may provide some service to a variety of other independent applications."

And one more comment on software from Open Projects:

> Software is created with programming languages and related utilities, which may come in several forms: single programs like script interpreters; packages containing a compiler, linker, and other tools; and large suites (often called integrated development environments) that include editors, debuggers, and other tools for multiple languages.

What this says, in so many words, is that software tools are often used to create applications and systems software. Software is also increasingly used to design and create hardware, especially integrated circuits. Therefore it is necessary to examine the security and integrity of hardware-creating software, as described by Villasenor [11].

However, before we take a more in-depth look at hardware, we need to understand the meaning of the term firmware (as shown in Figure 3.2). One definition is as follows [12]:

> ... firmware is ... often used to denote the fixed, usually rather small, programs and/or data structures that internally control various electronic devices. Typical examples of devices containing firmware range from end-user products such as remote controls ..., through computer parts and devices [such as] hard disks, keyboards ..., all the way to scientific instrumentation and industrial robots. Also more complex consumer devices, such as mobile phones, digital cameras ..., contain firmware to enable the device's basic operation as well as implementing higher-level functions.

It is interesting that both computer programs and data are included in the above definition of *firmware*. Some restrict firmware to program code and data that cannot be changed once burned into a chip. For example, some define *firmware* as unalterable software burned into read-only memory chips; this is the definition provided by the *IEEE Standard Glossary of Software Engineering Terminology* [13] defines *firmware* as follows:

> ... [t]he combination of a hardware device and computer instructions and data that reside as read-only software on that device.

However, today much firmware can be easily updated due to the ready availability of inexpensive nonvolatile memory devices. For example, many digital camera manufacturers provide updates to their firmware, which can be downloaded from the manufacturers' websites and installed on the appropriate cameras (as applied to medical equipment, scientific instrumentation, industrial control systems, and robots) opens up dangerous new possibilities for compromise. We note again that the category "data" is included in the IEEE definition. This is a more reasonable definition in the case of firmware, in contrast to application software because firmware updates will likely modify both functionality and data at the same time.

Wikipedia goes on to affirm that there "... are no strict boundaries between firmware and software, as both are quite loose descriptive terms. However, the term *firmware* was originally coined in order to contrast to higher level software which could be changed without replacing a hardware component, and firmware is typically involved with very basic low-level operations without which a device would be completely nonfunctional. Firmware is also a relative term, as most embedded devices contain firmware at more than one level."

The statement is also made that "while high-level firmware (or software) typically is stored as a configuration of charges, low-level firmware may instead often be regarded as actual hardware."

This presents a dilemma because developers of application-level software (as opposed to system-level software) tend to know little or nothing about the underlying hardware platform—they don't need to in order to fulfill their responsibilities. However, the November 2011 issue of the IEEE *Computer* magazine highlights the need for designers of software to work with hardware engineers [14]. This is because (particularly for exascale systems, such as computers with the capacity to handle 10 to the 18th power operations per second and other large-scale systems), the hardware is hitting against restrictions in cooling. Greater applications and systems software efficiency are needed in order to continue to expand the size of such systems. This calls for a heretofore unprecedented level of cooperation and collaboration. While the pressure on smaller-capacity systems is not as great, cloud computing has required a major increase in processing power and storage capacity to the extent that the availability of cheap electrical energy (often only available from hydro-electric powered generators) has become a major factor in determining the locations of cloud-computing data centers. There is clearly an increasing need for a holistic view with subject-matter experts from a number of different specialties being required to design and develop software-based systems in concert with one another.

Now let us take the Open Projects definition of hardware [15]:

> Harware is a comprehensive term for all of the physical parts of a computer, as distinguished from the data it contains or operates on, and the software that provides instructions for the hardware to accomplish tasks. The boundary between hardware and software is slightly blurry - firmware is software that is "built-in" to the hardware, but such firmware is usually the province of computer programmers and computer engineers in any case and not an issue that computer users need to concern themselves with.

The assertion that end users should not be concerned about the software-firmware-hardware mix is questionable because the specifics of such a mix not only affect computer and network performance, but also systems availability, security, resilience, and the like. If end users are confronted with choices with respect to the software-firmware-hardware mix, they may well opt for different configurations from those that hardware and software engineers would favor. These issues are generally worked out during the requirements phase, which is the phase of the development life cycle in which user needs are presented, matched against technical capabilities, and negotiated to arrive at the optimal balance between features and technology. The requirements phase is by far the most critical step in software systems engineering.

All of this is very confusing and can, to a large extent, be blamed for the lack of confidence in software engineering expressed by some researchers. The problems of understanding about software systems engineering often stems from the lack of accurate definitions. Software is not an amorphous generic blob, but rather a spectrum of capabilities covering software, firmware, and hardware. Increasingly, these specialties will need to talk to one another. It is hoped that the relatively disciplined structure of hardware systems engineering will be carried across to the software systems arena to the benefit of all stakeholders.

Are Control Software Systems Different?

We will now take a slight detour to ask the question as to whether software managing information systems and industrial control systems are different with respect to how we should approach them.

As we will discuss later, developers and users of many information systems have (until recently) had a much greater concern about security than those designing and developing control systems. High-integrity industrial control systems are needed to manage power plants, oil refineries, airplanes, weapons, and the like. Historically, these control systems have been isolated from other systems and public networks, and have been independent of one another and from business information systems. They also have tended to reside more on the hardware-firmware end of the spectrum that do information systems. However, controls systems are increasingly being networked and linked to information systems and networks that expose the said control systems to greater cyber security risks. The relatively new term for such systems is *cyber-physical systems*.

In earlier times, automated control systems could be considered as separate from information systems, with control-system emphasis solely on safety and with little or no attention paid to security. However, this distinction is becoming less and less appropriate as both types are combined into systems of systems [16], creating many new areas of vulnerability. Table 3.1 illustrates some of these areas of focus as they relate to information systems, control systems, and combined systems, respectively.

What is Software Systems Engineering?

Having negated the very existence of software systems engineering as a distinct field, we now look to define software engineering and software systems engineering as if they can be legitimized—which they can be. As shown in the

Table 3.1

Information Systems and Control Systems Versus Security and Safety

Focus	Information Systems	Control Systems	Combined Systems
Security	Loss of confidentiality, integrity or availability, leading to financial losses, fraud, etc.	Unauthorized access by persons or malware to applications that control critical systems, such as the compromise of isolated software managing centrifuges processing nuclear materials leading to their destruction.	Unauthorized access to networks and takeover of management of control systems, such as electricity generation plants and distribution grids, natural gas lines, refineries, waste treatment plants, weapons manufacturing, etc.
Safety	Potential physical damage to external subjects due to loss of confidentiality, integrity and availability (e.g., failure or compromise of a GPS system that might lead to capture of a military drone and top secret information).	Malfunctioning and/or faulty processes leading to inappropriate failure of control systems and subsequent risk to human wellbeing, such as for space vehicles, airplanes, automobiles, and the like.	Unexpected or planned destruction of facilities and/or injury to or death of humans from misbehaving control systems leading to compromise or failure of communications networks and systems

hierarchical diagram in Figure 2.2, software is considered a subcomponent of technology, with other subcomponents being made up of firmware, hardware, and networks. One might also say that software is involved with and (in some cases) committed to the creation and operation of firmware and equipment, where equipment might be systems, networks, and so on.

We now return to Braude and Bernstein [5], who define the purpose of software engineering as follows:

> The goal of software engineering ... is the creation of software systems that meet the needs of customers and are reliable, efficient, and maintainable. In addition the system should be produced in an economical fashion, meeting project schedules and budgets.

However, this raises the question as to why the field is not commonly called software systems engineering rather than software engineering. Is there a significant difference between the two terms? Yes, there is a big difference, and it is likely that this difference has contributed greatly to the largest problem in the creation of secure software. It has arguably led to incomplete requirements, inaccurate specifications, poor design, insecure code, inadequate testing, lack of metrics, and a whole litany of complaints about the vulnerabilities found in software.

The fundamental issue here is that software does not run in isolation. Any piece of software operates only in one particular context or in a limited number of environments. The context includes hardware, people, and processes.

End-user applications depend on software platforms, as well as the other applications and system components. Applications are designed to run on specific combinations of platforms and infrastructures (hardware, networks, etc.), and their security posture is highly dependent on the security of the environment in which they operate. For example, an application built for the iPhone may also run on an Android phone, but the application's security depends largely on how secure a specific phone might be at any given time.

We should do away with sloppy nomenclature such as software assurance, software engineering, and secure software engineering. We need to use terms such as software systems assurance, software systems engineering, and secure software systems engineering, which are much more specific and meaningful.

The Software Systems Engineering Process

The process for engineering software and related systems has been well-established for at least four or five decades. It is really surprising how so little has actually changed in the ensuing decades insofar as the software development life cycle is concerned.

Braude and Bernstein [5] and Lewallen [17] both give interesting histories of the software process, indicating the advantages and disadvantages of each method. The approaches and their positives and negatives are summarized in Table 3.2.

We also show the approximate year of origin of the various approaches. These dates are provided by the various sources referenced; they are only provided as a guide. Nevertheless, the chronology explains why certain authors omitted some approaches because, even if they had already been developed, they may not yet have been made public.

Steps in the Software Development Process

To some degree, all approaches contain the fundamental steps of the Waterfall approach, which contain the steps indicated in Table 3.3.

Royce is known as the person who first formalized the Waterfall approach in an article [18]. While Royce presented a linear model, it was only for illustration purposes. He propounded an iterative model later in his paper [18], but many people nevertheless inappropriately adopted the linear model. That Royce himself did not advocate the linear model and supported the iterative model is discussed in more detail in [19].

The phases of the SSDLC, as listed in the first column of Table 3.3, were common in the era of mainframe computers when computer systems, along with developers and operators, were isolated with a few (if any) remote users connected via networks. The software development and implementations processes adhered to fairly consistent, structural lines and followed standard systems engineering practices and procedures. Systems analysts obtained user requirements, analyzed the inputs, and produced software system designs or specifications. These designs and specifications were then developed and tested by programmers, tested in the quality assurance area, and were then turned over to production. There were only a few platforms available, namely, IBM, Digital Equipment Corporation (DEC), Data General (DG), and a few other special-ized systems, such as Control Data. IBM had a few operating systems, and vari-ations of the UNIX operating system were commonly used on minicomputers. Maintenance was usually performed by the developers/programmers, and en-hancements required that systems analysts update requirements and go through the same steps as for the original development process. Documentation relating to systems requirements and design was generally provided by systems analysts, and programmers were given the responsibility of generating documentation about the code that they had written, so that they could more readily maintain and upgrade the system. Operational and user manuals were also written by systems analysts closest to each aspect.

Because the data processing departments of that earlier era were both limited and isolated, security was generally implemented by physical meth-ods; this is because users were often bound by the need for physical proximity. Consequently, it was much more feasible than it is today to secure central and remote physical locations and then implement an access management system on the mainframe computers. Smaller departmental computers, often in the form of minicomputers running the Unix operating system, could be physically secured. However, software-based access controls were much less common for microcomputers than for mainframes.

In some of the approaches listed in Table 3.2, processes address each step sequentially, in others there is feedback from one step to another, and in still others there are iterations of the entire process. Also, individual approaches tend to stress particular steps and play down others. Preferences among approaches often depend on the roles and duties of a particular stakeholder. Thus, a devel-oper will likely prefer the Agile approach which deemphasizes documentation, whereas a support person would favor an approach that yields specific and ex-tensive documentation.

The texts referenced go into much greater detail with respect to the vari-ous methods, and guidance is provided as to which method might be the most appropriate for a given situation. For example, if the project is relatively small and contained, one might favor the Waterfall approach and shy away from the Spiral approach.

Table 3.2
Advantages and Disadvantages of Various Software Development Process Approaches

Approach	Year of Origin*	Major Advantages	Major Disadvantages
Waterfall	1970	Simple to use Large population of experienced project managers Easy to manage Phases are processed and completed one at a time Facilitates resource allocation Good for smaller projects	Need to know full requirements up front (adjusting scope during the life cycle can kill project) Many problems not discovered until late in the process at which point working software is produced Lack of parallel activities Relatively inefficient use of resources High risk and uncertainty particularly if requirements are likely to change Not good for complex and object-oriented projects Not good for larger projects that are long-lasting and ongoing
Spiral	1988	Planning built into process Risks addressed early and on continuing basis Software produced early and evolves throughout process Good for large and mission-critical systems	Complicated and costly Risk analysis requires very specific expertise Success of project depends very much on the quality of the risk analysis Overkill for small projects
V-Model	1986	Easy to use There are specific deliverables required at the end of each phase Test plans are developed early in the life cycle thereby reducing risk Good for smaller projects with well-defined, easy-to-understand requirements	Not flexible Changing scope is difficult and expensive No early prototypes are produced Not clear how problems found during testing are handled
Iterative Incremental	1993	Facilitates setting of clear and manageable objectives Reduces overall project complexity and risk Better specification of requirements by users Less costly to change scope and requirements than with some other approaches Iterations are more easily managed Problems are identified and handled at each iteration rather than waiting to the later stages	Greater level of coordination required among subprojects Each phase of an iteration is rigid and inflexible Phases within an iteration do not overlap Potential problems with system architecture because of inadequate requirements during early stages

Table 3.2 (continued)

Approach	Year of Origin*	Major Advantages	Major Disadvantages
Prototyping Proof of concept	1983	Facilitates understanding of factors, such as look and feel of user interface Helps to identify and mitigate risks May be able to reuse some of prototype code Avoids the cost and difficulty of modifying completed software	May not be cost-effective in certain cases
Unified	1999	Comprehensive approach Includes aspects not included in other methods such as business case Quite mature and widely used	Overkill for small projects
Agile	2001	Demonstrable results Developers may be more motivated Better specification of requirements by users	Problematic for larger projects Documentation suffers
Open source		Utilization of potentially large number of resources Encourages personal satisfaction Allows for tailoring and integration More stable maintenance	Little to no control over requirements Inconsistent or poor documentation Little or no assurance of availability of developers and support personnel Lack of management control

*Approximate

Table 3.3
Phases of the SSDLC

Royce [18] (1970)	Christensen [6] (2001)	Braude [5] (2011)
System requirements	Requirements	Requirements
Software requirements	—	—
Analysis	—	—
Program design	Design	Design
Coding	Code and unit test	Implementation
Testing	Test and integration	Testing
Operations	Operation and maintenance	Maintenance

SSDLC = software system development life cycle.

Omissions or Lack of Attention

Research into the various sources relating the software systems engineering suggested that there might be several highly important omissions in all of the approaches. These omissions might account for many of the continuing problems emanating from software. We mention these deficiencies below and expand upon them in subsequent chapters.

Nonfunctional Requirements

It is particularly interesting how certain so-called nonfunctional attributes of software systems are systematically neglected in publications on software engineering. These requirements, which include performance, security, safety, dependability, resiliency and the like, are termed "nonfunctional" because they are not usually included in the functional requirements of software as specified by users. They are considered to be the responsibility of other areas of expertise, such as system and network capacity planners, information security professionals, software safety engineers, disaster recovery and business continuity planners, and so on. Unfortunately, this attitude of nonfunctional requirements being in the realms of others leads to situations where nobody takes on the responsibility.

While these nonfunctional requirements are always key attributes of any substantial software system, the average application developer is not particularly knowledgeable in these nonfunctional areas and therefore tends to ignore them. In order to have nonfunctional requirements included, project managers must account for these characteristics explicitly in the various stages of the software system development life cycle. Otherwise, nonfunctional requirements will tend to be ignored—sometimes until it is too late to save the project [20].

Testing Nonfunctional Attributes

Braude and Bernstein [5] state that a major limitation of the Waterfall process is that the testing phase sits at the end of the cycle; therefore, errors that might have been determined with interim testing are only discovered when the software has been completed. They also affirm that it is only at the point when the project is complete that "[m]ajor issues such as timing, performance, storage, and so on can be discovered ..." Interestingly, neither security nor safety are mentioned in their list of issues.

Braude and Bernstein [5] seem to imply that the original model, as proposed by Royce [18], was unidirectional, when in fact Royce specifically stated that the process should be iterative between immediately adjacent steps. Braude and Bernstein [5] state that, practically speaking, "... an iterative model between successive phases is inevitable." As mentioned previously, there is an ongoing debate about whether the simple Waterfall model, as adopted by the United States and European military and others, was due to a lack of understanding of these adopters. Clearly, Royce did not advocate the simple model, but insisted on the necessity of including interactive and iterative aspects into the process model.

Verification and Validation

It is not clear when the term *verification and validation* was first introduced, although there is some indication that Barry Boehm was an early user of the term around 1981. The so-called "testing" of software usually involves the following series of activities:

- Unit testing;
- Integration testing;
- Regression testing;
- User acceptance testing.

These activities are all basically related to functional testing and are meant to ensure that a piece of software operates successfully in isolation, when combined with and connected to other components of the system, and when operated in its most likely overall production environment, and that the functionality as developed is acceptable to the user community. Other areas, such as technical support, user support, and operations, are usually included to some relatively small degree in such testing activities to determine that the system also satisfies their particular requirements. Those responsible for nonfunctional areas, such as performance, resiliency, safety (to a lesser extent), and security, may well be excluded from the system development life cycle process until the final stages, when the software is essentially completed and ready to go into production. At this point, it is either too late or extremely expensive to make requisite changes. It is commonly held that it is orders of magnitude more expensive to implement security, say, after the system has been completed compared to building in security from the earliest stages of the life cycle.

On the other hand, verification and validation takes a more holistic view of the testing and quality assurance functions.

The IEEE definitions of these terms, along with clarifying comments are included in [5] as follows:

- *Verification:* "The process of evaluating a system or component to determine whether the products of a given development phase satisfy the conditions imposed at the start of that phase." (For example, is the software design sufficient to implement the previously specified requirements? Does the code correctly implement the design?)

- *Validation:* "The process of evaluating a system or component during or at the end of the development process to determine whether it satisfies specified requirements." (In other words, does the implemented system meet the specified user requirements?)

These definitions are summarized in [5] as follows:

- *Verification:* Ensuring that each artifact is built in accordance with its specifications. (Mostly inspections and reviews: "Are we building the product right?").

- *Validation:* Checking that each completed artifact satisfies its specifications (Mostly testing: "Are we building the right product?").

These definitions are somewhat reminiscent of the definitions of efficiency and effectiveness, respectively; where efficiency is defined as doing things the right or correct way (comparable to verification), and effectiveness is defined as doing the right or correct things (comparable to validation).

In effect, what is being said here is that verification determines that the resulting system or component represents requirements (as specified by the software engineers who designed the system or component), whether or not those requirements accurately reflect what the user population is looking to get from the system. Validation, on other hand, determines whether in fact user requirements are being met. If either of these tests show that the system does not represent either system specifications or users' expectations, one has then to determine whether the gap is due to improperly translating user requirements into system specifications and into the design and development of the system or component, or whether the difference is the result of misinterpreting users' expectations of the system or component. This difference is illustrated in Figure 3.3.

The V-model, which is shown in Figure 3.4, captures all the phases of the waterfall model and includes reviews and testing consistent with the verification and validation processes. However, it still has the same issues as the waterfall model with respect to rigidity and inflexibility.

This difference between efficiency and effectiveness (verification and validation, respectively) is also illustrated by such approaches as the Capability Maturity Model Integration (CMMI) process. While CMMI can demonstrate whether or not an organization has effective software system development processes, procedures, and practices in place, it does not ensure that the appropriate software is developed, despite the fact that the software was developed according to a highly structured process that is being monitored and measured at every step. One might presume that—if a software product was manufactured

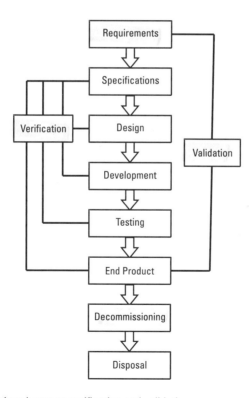

Figure 3.3 Comparison between verification and validation.

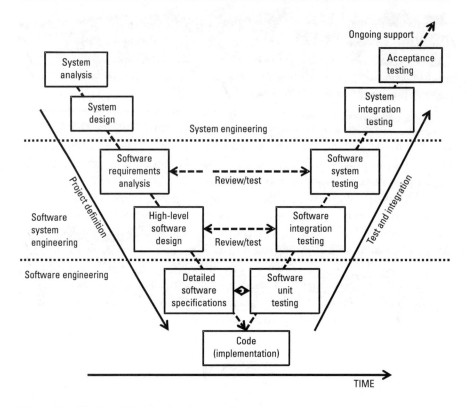

Figure 3.4 The V-model of the development life cycle process.

using agreed-upon good practices—it is likely that such software will also meet user requirements. However, that is not necessarily the case. In fact, even the most sophisticated of processes tends to omit whole categories of testing, and applications are frequently designed without having the ability to generate key security data.

Creating Requisite Functional and Nonfunctional Data

There is a common tendency towards working with readily-available data to create metrics by which to manage and control what gets built into software and what does not, as well as how the software is created. This assertion may be less true when measuring software functionality than it is for measuring security and other nonfunctional requirements. In [21], Axelrod takes a relatively extreme position, stating that:

> … measurability and usefulness are in contention when it comes to security metrics.

This relationship is illustrated in Table 3.4, in which we show the relationship between the value or usefulness of measured information and the difficulty and cost of creating such data.

Sometimes very useful information can be gathered very cheaply, but that is not the general rule. More often than not, easily-obtainable data and readily-calculated metrics are of relatively little value to decision-makers. Such easy-to-get metrics may point out gross deficiencies or vulnerabilities, but are often of little use in finding the root cause of a problem. Occasionally, readily-available data leads right to the problem, but even then it may have to be supplemented with secondary information before a full forensic analysis can be done. Thus, if you lose a network connection, it might be easily explained by, for example, a router having become unplugged. However, if the loss of network connection is due to malware, for example, unless you have effective monitoring and reporting systems in place, you may not be able to detect the real reason for the outage.

At the other extreme, you might invest in a sophisticated detection system either built into the system or an add-on, in order to obtain information on exactly what is causing a particular set of problems. Intrusion detection and prevention systems come to mind in this case, particularly if they have the capability to detect nefarious behavior. From a software perspective, much can be achieved by incorporating data creation, logging, and reporting requirements into applications and system software during development.

Table 3.4 presents the idea that it is more likely that easily and cheaply obtained information is of relatively little use for decision-making, as compared to data that are prescribed ahead of time but which might require more effort and overhead to produce and analyze. Also, there is a danger of investing large amounts of money and effort into creating data that are not very useful; it is important to think through data requirements carefully and have them reviewed by experts in order to avoid wasting money on an approach that yields little in the way of useful information.

Generating as much information, even useful information, as possible is not always cost-effective, as it will be subject to diminishing returns on investment. There is usually a point up to which such investments in security and

Table 3.4
Measurability Versus Usefulness of Data

	Easy/Cheap to Measure	Difficult/Expensive to Measure
Highly Useful	Preferred situation, but relatively rare occurrence	Often the case, but results likely to yield substantial benefits
Not Useful	Unproductive, but more likely occurrence	Wasteful metrics gathering programs with little to show for significant expenditures and effort

safety tools are warranted and beyond which they are not. Because the cost of incorporating data generators and collectors is considerably less expensive if introduced early in the software system development life cycle (SSDLC) than if added as an afterthought when the software has already been manufactured and tested, the cost-benefit analysis of generating and collecting relevant data will produce very different results when the data generators and collectors are built in than if they are added after the system has been completed. We show this relationship in Figure 3.5, namely, full incorporation of data generation, collection, analysis, and reporting at all stages of the development life cycle versus bolting on whatever is possible once the system has been essentially completed.

In Figure 3.5, we show a range of costs and values of incorporating data generating and collection functions as we progress through the life cycle. The graphs indicate that the range of potential costs and values widens with later steps in the life cycle because there is greater uncertainty as to what the costs might be and what benefits might be derived from the added capabilities. In this example, an average net value is shown to turn from positive to negative in the later stages of the SSDLC. This means that, at some point, it no longer pays to include such data creation capabilities because they would cost more that might be derived from them, with the implication that you would just have to bear the risk and costs of not being able to monitor certain activities. However, the graph also shows that, if the data creation and monitoring are considered and included at the requirements stage or soon thereafter, the costs will be

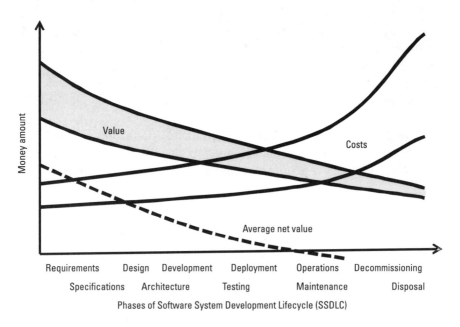

Figure 3.5   Costs of measurability and value of metrics for phases of the SSDLC.

relatively low, and the benefits of being able to include considerable data generating, logging, and reporting capabilities significantly exceed the costs. While this particular example shows that at some point it is not worthwhile to retrofit the data creation capabilities, in some cases, the value of being able to monitor behavior may be so high that it might still be beneficial to bolt the capabilities on after the system is already in production. The same is true for security, safety, and many other nonfunctional attributes. That is, there might be a stage where it is no longer cost-effective to incorporate security features, say, in which case the risk has to be assumed.

The obvious conclusion here is that, by incorporating data generation, collection, and analysis early in the process, lower relative costs are incurred and, as a consequence, more data generation and collection can be justified leading to greater net value. However, if one waits too long, then it is no longer cost-effective to build the capabilities in and the previously available benefits of doing so are lost.

Resiliency and Availability

A feature of software that frequently does not receive the attention that it requires is that of resiliency. There are likely many reasons for this lack of attention, but one reason is surely because there are many different ways in which to provide resiliency and many are outside the usual purview of the average software engineer. Cross-discipline and cross-cultural factors play a large part in many aspects of engineering safe and secure software systems, but perhaps none more so than in the resiliency area. There is a need to optimize across software, platforms, infrastructure, and facilities, and few individuals have the knowledge and experience to deal with every aspect. Therefore, it becomes necessary to establish collaborative teams made up of a range of disciplines in order to get the level of coverage needed.

The main issue with respect to software resiliency is that it is multifaceted. Some ways in which resiliency can be achieved are inherent in the design of software, others relate to the platforms on which the software runs and the infrastructures that support the platforms. Others have to do with physical redundancy both on-site and off-site.

Again, the issues revolve around whether resiliency was included as a requirement and incorporated into the software system specifications. If not, it is likely that the costs of providing redundancy will be significantly greater than they would have been if the issue is ignored until after the system has been developed, tested, and deployed. This can also be represented by the graphs in Figure 3.5, although the underlying justification is considerably different. In the case of resiliency, there are many choices (software, firmware, hardware, platform, infrastructure) to achieve adequate resiliency, only some of which relate to the application software, but many of which can be handled externally to the software. This is not the case with respect to data creation.

Decommissioning

Few approaches to the software system development life cycle process address the discontinuation and decommissioning of the software. Despite the availability of regular updates for commercial software, at some point the software becomes obsolete and is no longer supported by the software maker. At that point, a number of issues arise. One is the concern that new functionality will not be available for this particular software product. Another is that newly discovered errors will not be corrected. And yet another is the concern that patches will no longer be issued for newly discovered vulnerabilities [22].

However, at some point it is inevitable that obsolete software products will have to be replaced for such reasons as lack of functionality, lack of support, need to interface with newer software and services, and so on. The question then arises as to what to do with the obsolete software and its attendant data. The issue seems to have been given more attention lately because of privacy and piracy issues, for which concern arose as to how sensitive data were being disposed of. In some cases, the data became integrated with applications and it was not feasible to separate them out for destruction and disposal. The decommissioning and disposal of software systems needs to be accounted for early in the life cycle in order to ensure that these important activities are considered at all.

Summary and Conclusions

In this chapter, we considered what constitutes software systems engineering, which is considered by some to be a major source of deficiencies in modern information-technology and automated control systems. This proved to be a nontrivial exercise. As was demonstrated, much of the criticism surrounding software engineering is justified and can be attributed to deficiencies in the definition of terms and the confusion surrounding scope and completeness. By offering more specific definitions and formal structures, we have attempted to clarify the meaning and scope of the field of software systems engineering.

Another major area to be considered relates to gaps in the traditional software development life cycle, particularly as they relate to scope, testing, resiliency, decommissioning, disposal, and the like. These gaps can be blamed, to a significant degree, for vulnerabilities and other weaknesses in software. Many of the ideas developed in this chapter will be expanded in subsequent chapters.

Endnotes

[1]   Pesactore, J., "No Insurance Policy Ever Protected a Customer, and Lots of them Don't Even Limit Business Risk," *Gartner Blog*, July 22, 2011, at http://blogs.gartner.com/john_pescatore/2011/07/22/no-insurance-policy-ever-protected-a-customer-and-lots-of-them-dont-even-limit-business-risk/.

[2]   Parnas, D. L. "Software Engineering—Missing in Action: A Personal Perspective," *Computer*, Vol. 44, No.10, October 2011, pp. 54–58.

[3]   Axelrod, C. W., "Does FOSS Pay? Weighing the Security Risks and Benefits of Open Source Software," *ISSA Journal*, July 2006, pp. 6–12.

[4]   Davis, M., "Will Software Engineering Ever Be Engineering," *Communications of the ACM*, Vol. 54, No. 11, 2011, pp. 32–34.

[5]   Braude, E. J. and M. E. Bernstein. *Software Engineering: Modern Approaches, Second Edition*, Hoboken, NJ: John Wiley & Sons, 2011.

[6]   Christensen, M. J. and R. H. Thayer. *The Project Manager's Guide to Software Engineering's Best Practices*, Los Alamitos, CA: IEEE Computer Society, 2001.

[7]   Allman, E., "The Robustness Principle Reconsidered," *Communications of the ACM*, Vol. 54, No. 8, 2011, pp. 40–45.

[8]   Definition of *software*, accessed at http://www.openprojects.org/software-definition.htm on July 11, 2012.

[9]   Definition of *systems software*, accessed at http://www.openprojects.org/software-definition.htm on July 11, 2012.

[10]   Definition of *application software*, accessed at http://www.openprojects.org/software-definition.htm on July 11, 2012.

[11]   Villasenor, J. "The Hacker in Your Hardware: The Next Security Threat," *Scientific American*, Vol. 303, No. 2, August 2010.

[12]   Definition of *firmware*, accessed at http://en.wikipedia.org/wiki/Firmware on July 11, 2012.

[13]   *IEEE Standard Glossary of Software Engineering Terminology*, STD 610.12-1990.

[14]   Shalf, J., D. Quinlan, and C. Janssen, "Rethinking Hardware-Software Codesign for Exascale Systems," *Computer*, Vol. 44, No 11, November 2011, pp. 22–30.

[15]   Definition of *hardware*, accessed at http://www.openprojects.org/hardware-definition.htm on July 11, 2012.

[16]   In recent times, researchers have been using the term *systems of systems*, which is defined in many different ways. According to M. W. Maier, in his article, "Architecting Principles for Systems-of-Systems," accessed at http://www.infoed.com/Open/PAPERS/systems.htm , on December 18, 2011, "Systems-of-systems should be distinguished from large but monolithic systems by the independence of their components, their evolutionary nature, emergent behaviors, and a geographic extent that limits the interaction of their components to information exchange. Within these properties are further subdivisions. For example a distinction between systems which are organized and managed to express

particular functions, and those in which desired behaviors must emerge through voluntary and collaborative interaction."

[17]   Lewallen, R., "Software Development Life Cycle Models," July 2005. Accessed on July 11, 2012 at http://codebetter.com/raymondlewallen/2005/07/13/software-development-life-cycle-models/

[18]   Royce, W. W., "Managing the Development of Large Software Systems," *Proceedings of the IEEE WESCON*, 1970. Reprinted article, http://www.cs.umd.edu/class/spring2003/cmsc838p/Process/waterfall.pdf.

[19]   htt://tarmo.fi/blog/2005/09/don't-draw-diagram-of-wrong-practices-or-why-people-still-believe-in-the-waterfall-model/.

[20]   The author is familiar with a system that was designed and developed by bright, but somewhat inexperienced, software engineers who concentrated on the graphical user interface (GUI) to the exclusion of other system characteristics. After an effort of many years and costs of perhaps $80 million (the exact amount was never publicized), the new system was placed side-by-side with the "green screen" legacy system in order to compare performance. The older system responded to an inquiry within 3 seconds. The new system took about 30 seconds to respond to the same inquiry. After unsuccessful attempts to improve the performance, the new system was abandoned, eventually to be replaced by an entirely different system.

[21]   Axelrod, C. W. "Accounting for Value and Uncertainty in Security Metrics," *ISACA Journal*, Vol. 6, 2008.

[22]   As an example, when Microsoft decided to stop supporting Windows NT 4.0, there was uproar among customers who had not upgraded their software within the warning period. There was particular concern about the possibility of security-related vulnerabilities appearing and not being addressed. Microsoft agreed to have customers sign up for a special extended support service dedicated to security patches. Initially, a steep fee was attached to this service; it is possible that Microsoft ended up not charging for the service.

4

Engineering Secure and Safe Systems, Part I

Security engineering is about building systems to remain dependable in the face of malice, error, or mischance.
—Ross J. Anderson, Cambridge University

Introduction

When it comes to building, operating, and decommissioning safe and secure systems—particularly computer-based software-intensive systems—practitioners often borrow methods and practices derived from the discipline of systems engineering, as well as from safety and security engineering. In this chapter, we take the more general systems view of engineering secure and safe systems. This is somewhat more difficult than one might think because so much of what has been published recently, while having a software orientation, it does not adequately treat specific security and safety attributes of the systems under discussion.

This assertion about the predominance of software considerations is reasonable because practically all modern systems of any complexity whatsoever incorporate software components to some degree. However, for the sake of clarity, we defer consideration of the engineering of safe and secure software-intensive systems until later. In this chapter, we will focus on safety and security considerations for general systems before making everything that much more specific and complicated by adding the software dimension.

The Approach

Just as it was recommended in Chapter 3 that the term software engineering be expanded to software systems engineering in order to account for the fact that software products only run in specific contexts, so security engineering and safety engineering need to be more clearly identified and defined so that it is clear as to what is meant by security and safety as they apply to systems engineering, where systems may or may not include computers. In this chapter, we delve into the specific areas of security engineering and safety engineering. In subsequent chapters, we will relate security and safety to software-intensive systems.

We now go through a similar defining process as the one followed in Chapters 2 and 3 for clarifying what is meant by security engineering and safety engineering. We thereby set the stage for determining what it takes to apply these concepts to software systems, as we will do in Chapter 5. It is again asserted here, as it was in earlier chapters, that being more precise in defining such terms is the key to better understanding and should be helpful in shedding light on much of the fuzzy thinking that we find in this area. Many prominent researchers and authors do not take the trouble to differentiate between security and safety engineering, or perhaps they do not recognize that there might be a difference. They also appear not to care to make any particular distinction among security, safety and other descriptors, such as dependability. This makes it considerably more difficult to make comparisons across disciplines. Here, we try to overcome much of this confusion and lack of precision by separating out the terms and examining each one separately and then in combination with themselves and also with other terms.

Security Versus Safety

In his book on the subject of security engineering, Anderson [1] certainly covers practically every conceivable topic that one might associate with security engineering as it relates to software systems. However, his distinction between the concepts of security and safety is not particularly clear. This is not uncommon. Many writers use the words security and safety interchangeably. Here, we will differentiate carefully between the two terms. There is a considerable gap in knowledge and communications between those who focus on security and those specializing in safety, whether referring to general systems or software-intensive systems, so that it is still appropriate to treat the disciplines independently. It is hoped that the separation between the safety and security engineering researchers, writers, and practitioners will lessen over time as participants will be called upon to address both safety and security issues [2].

As discussed in Axelrod [3], there is an increasing need to identify the frequently-separated silos of security and safety. It is also important to bring security and safety professionals together to collaborate at each stage of the system development, testing, operation, and decommissioning life cycle per Carter [4]. This is particularly urgent, as we are increasingly seeing safety-critical and security-critical systems linked together to form systems of systems. The so-called smart grid, in which electricity power grids' management systems are increasingly being linked to public networks, particularly the Internet, is a current case in point [5]. The apparent confusion and lack of understanding by management and technologists underline the importance of differentiating between expertise relating to safety and security, and attaining a greatly increased measure of cross-cultural cooperation, as recommended by Carter [4].

A simple, yet meaningful, distinction between safety-critical and security-critical software systems, which is generally applicable to system was provided in Chapter 2 (p. 15) for software. Boehm's original definition was repeated in Barnes [6]. We generalize the definitions as follows:

Safety – the [system] must not harm the world,
Security – the world must not harm the system.

This difference in viewpoint is illustrated graphically in Figure 4.1 for safety-critical, security-critical, and combined safety/security-critical systems, respectively.

Briefly, the goal of safety-critical systems is containment, whereas the corresponding emphasis of security-critical systems is protection. Such a difference in objectives leads to varying and sometimes conflicting criteria for ensuring that systems are safe when compared with those criteria used to ensure security.

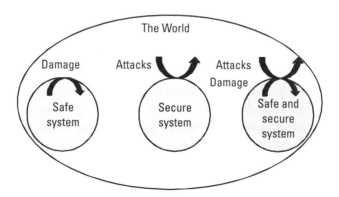

Figure 4.1   Safety versus security.

There is overlap to some extent between security and safety, as, for example, in the need for system integrity, but it is the difference in viewpoint that leads to divergent system designs and implementations. This divergence makes for much higher risk relating to combined systems of systems which must be both secure and safe.

Another perspective derives from the differences among safety, security and survivability, as described by Firesmith in [7]. Firesmith claims that there are many similarities among these three aspects, which he defines as follows:

- *Safety*: The degree to which *accidental harm* is prevented, detected, and reacted to.

- *Security*: The degree to which *malicious harm* is prevented, detected, and reacted to.

- *Survivability*: The degree to which *both accidental and malicious harm* is prevented, detected, and reacted to.

Firesmith goes on to say that "[s]afety is largely about protecting valuable assets (especially people) from harm due to accidents," whereas "[s]ecurity is largely about protecting valuable assets (especially sensitive data) from harm due to attacks," and "[s]urvivability is largely about protecting valuable assets (essential services) from both accidents and attacks."

This author prefers the Barnes approach [6] to differentiating between safety and security, because from the perspective of those tasked with protecting the data assets contained within, or accessible through, information systems or preventing control systems from harming the outside world, it often doesn't make much difference, with respect to actions taken, whether the source of an incident is accidental or intentional and malicious. In both situations, investigators need to determine the root cause of an incident and take action to prevent the same or similar future events from doing damage, either internally or externally. However, if an incident is the result of intentional activities, the response to it will likely be different in that law enforcement might be brought in to pursue the perpetrators. If the cause is accidental, action against the person or persons who "allowed" the incident to happen may result, even if such persons had no intent or control over the situation. However, in the latter case, it is less likely that formal proceedings will be invoked.

There is often a question as to the culpability of individuals and the fairness of pointing fingers at suspects. Frequently, even if the cause of a security breach or a safety failure is found to be unintentional, action is taken against those currently responsible for the safe and/or secure operation of the affected system, despite the fact that they may not be in a position to prevent the incident. The root cause might have been deficiencies in requirements, design or implementation of systems or the procedures for operating them. However, it seems to be human nature to seek a scapegoat on whom to pile responsibility and punishment.

It can be argued that survivability has more to do with resiliency and recovery following damage or destruction, and long-term survivability relates to the ability of an organization to reconstitute the damaged or destroyed operations in order to enable the entity to function at or near the level prior to the incident.

Yet another view on differentiating between safe and secure software systems comes from Barbacci [8], who represented the same institution as Firesmith, namely the renowned Software Engineering Institute of Carnegie Mellon University. Both researchers were writing in 2003.

Barbacci quotes Rushby [9] in defining secure systems as "... those that can be trusted to keep secrets and safeguard privacy," and, paraphrasing Laprie's definition of dependability, defines safety as "... that property of a computer system such that reliance can justifiably be placed in *the absence of accidents.*"

Barbacci differentiates between safety and dependability as follows:

> ... dependability is concerned with the occurrence of failures, defined in terms of internal consequences (services are not provided) ... safety is concerned with the occurrence of accidents or mishaps, defined in terms of external consequences (accidents happen).

Barbacci concentrates on the outcomes of incidents and not on the causes. He differentiates between *internal* and *external* consequences; external consequences align with Barnes' definition of safe systems. Barbacci's definition of internal consequences actually focuses on the impact of a failure on the outside world, not in terms of damaging the outside world actively, but inconveniencing either internal or external users of the system or both, passively. That is to say, dependability is about the absence of regular, expected, trustworthy services and does not appear to relate to the compromise of data assets within information systems or the interference with normal operation of control systems.

If we reexamine the quote from Ross Anderson at the beginning of this chapter, it appears that security and dependability have more commonality than do safety and dependability. However, this view might be attributed to Anderson not having dealt much with safety issues in his book.

Four Approaches to Developing Critical Systems

Rushby [9] presents four traditional approaches to developing critical systems, namely:

1. The Dependability Approach
2. The Safety Engineering Approach
3. The Secure Systems Approach
4. The Real-Time Systems Approach

It is worth examining these four approaches to developing critical systems in greater detail because they introduce some of the key aspects of the subsequent treatment of software systems engineering later in this book.

The Dependability Approach

Rushby [9] asserts that the use of the term dependability is an attempt to free researchers from having to use other terms, such as reliability, which have acquired particular technical meaning. He claims that the term dependability can be used to include fault tolerance, reliability, correctness, safety, survivability, and security. Rushby appears to consider security and safety to be particularly significant factors in system dependability because they are the focus of two of his four approaches to critical system development.

Quoting Laprie [10], Rushby asserts that a dependable system is one upon which certain aspects of quality of service, such as correctness and continuity of delivery, can be relied. The dependability approach itself tends to emphasize reliability and fault tolerance, rather than safety or security.

In this book, we pay particular attention to the security and safety of systems, especially software-intensive systems, and we do not consider reliability and fault tolerance much. However, that is not to say that reliability and resiliency are not crucial in the development and running of safe and secure software systems. They are.

In Table 4.1, we show which dependability components are addressed by various researchers.

Referring back to the description of the dependability approach, it should be noted that Rushby states that a "departure from the service required of a system is [considered to be] a failure" [9], and that the implication is that dependable systems do not fail. However, there is a range of failures ranging from benign to catastrophic, where the consequences are hugely greater than benefits derivable from the system. Failure is considered a "property of the external behavior of a system" [9].

The causes of failures are called *faults* and include the following:

• Mistakes in design and implementation;
• Component failures;

Table 4.1
Components of Dependability by Various Researchers

Component	Anderson [1]	Barbacci [8]	Rushby [9]	Laprie [10]
Security	Yes	—	Yes	—
Safety	Yes	—	Yes	—
Correctness	—	—	Yes	Yes
Continuity of delivery	—	Yes	—	Yes
Reliability	Yes		Yes	—
Fault tolerance	Yes	Yes	Yes	—
Survivability	—	—	Yes	—
Quality of service	—	—	—	Yes

• Improper operation;

• Environmental anomalies.

According to Rushby, a system is fault-tolerant if it can detect and correct *latent errors*, which occur when a system is damaged by a fault.

The Safety Engineering Approach

Rushby [9] states that safety engineers distinguish between reliability and safety, noting that reliability relates to "the incidence of failures," whereas safety has to do with "the occurrence of ... mishaps." Mishaps are defined as *unplanned events* that might lead to "death, injury, illness, damage to or loss of property, or environmental harm." Again, this raises the question as to whether a failure brought about by a *planned event* should not also be included among failures. It should.

Leveson [11] proposes that those working on software safety adopt and follow some of the practices and techniques of system safety engineering, with a focus on consequences rather than requirements because omissions or incorrectness in requirements may well lead to negative consequences. In a subsequent chapter, we will critically examine the role of requirements in defining safety and security needs. This is because many failures and successful attacks that do occur could well be the result of correctly representing inadequate requirements. Rushby asserts that the dependability approach works best in situations where there "is no safe alternative to normal service," giving the example of aircraft flight control systems, whereas the safety engineering approach applies to situations in which hugely undesirable events, such as the inadvertent release of weapons, can happen.

Rushby defines hazards, damage, danger, severity and risk as follows:

- *Hazard:* conditions that can lead to a mishap;
- *Damage:* a measure of the loss in a mishap;
- *Danger:* probability of a hazard leading to a mishap;
- *Severity of a hazard:* the worst possible damage that could result;
- *Risk:* combination of hazard severity and danger.

Rushby goes on to describe hazard analysis, which, he asserts, is a major tool used in system safety engineering. This and the above terms appear to be specific to safety engineering, but they have their equivalence in the more general area of security risk analysis, as shown in Table 4.2.

According to Rushby, hazard analysis comprises the following steps:

1. Identification of potential hazards (ranging from *negligible* to *catastrophic*).

2. Systematic exploration, which is the determination of how or whether conditions leading to a mishap might arise (backwards reasoning and what-if analysis).

3. Dealing with unacceptable risks by means of modification of systems and operating procedures, incorporation of safety features and/or training.

Rushby contends that the use of hazard analysis should be required at each phase of the design life cycle, and he gives examples such as preliminary subsystems, systems and operations. The following methods are commonly used for hazard analysis:

- Hazard and operability studies (HAZOPS);

Table 4.2
Safety-Engineering Terms Compared to Security-Risk Terms

Safety-Engineering Term (Rushby [9])	Equivalent Security-Risk Term
Hazard	Threat, exploit, vulnerability
Damage	Magnitude of loss
Danger	Probability of loss
Severity of a hazard	Highest potential loss from threats and vulnerabilities combined
Risk	Risk, expected loss (magnitude × probability of loss)
Hazard analysis	Risk analysis

- Fault-tree analysis (FTA);
- Failure modes and effects analysis (FMEA).

These techniques were recommended for use with software by Leveson and Harvey [11] in 1983, specifically using software fault tree analysis (SFTA); the SFTA technique was developed in the late 1960s to avoid accidental launching of missiles.

Rushby states that the entire system context in which the software operates has to be considered when using the safety-engineering approach because software by itself might not reveal system-level mishaps. Rushby asserts that not only can a software error cause failure in the overall system, but that activities in the rest of the system can also cause the software to fail.

Rushby's article further describes thes afety system engineering implementation mechanisms, which involve lockins, lockouts, and interlocks that lock systems into safe states, lock them out of hazardous states, and require specific event sequencing, respectively. The determination of lockins and lockouts will vary with the criteria being used to select the appropriate failure mode. Missteps in this analysis can lead to horrific consequences, as when workers were locked into radioactive rooms at the Fukushima Daichi nuclear plant following the March 11, 2011 earthquake and tsunami hitting the northeast coast of Japan [12].

Rushby also mentions that safety approaches can be applied to security if mishaps are interpreted as unauthorized disclosure of information and threats, vulnerabilities, safeguards, and it countermeasures are considered.

The Secure Systems Approach

Rushby [9] claims that the main concern regarding system security has traditionally been the disclosure of sensitive information, with the protection of information integrity and the prevention of denial of service being secondary concerns. He describes security models as comprising both a system component and security component. The list of security models, which includes the Bell-La Padula access-control model, differs among system components, such as sequential systems, distributed systems, databases or expert systems, and for security components.

Rushby notes that the disclosure of sensitive information usually receives more attention than information integrity, even though integrity may be considered to be more important in some cases because its loss might have greater impact on the overall system than would compromise of confidentiality. He refers to Clark and Wilson [13], who focus on two aspects of integrity, namely, well-formed transactions and separation of duties. They describe well-formed transactions as those that require that "important data cannot be modified by arbitrary actions," and *separation of duties* as requiring that "different individuals authorize the different procedures that constitute a larger action." Separation of duties should be across departmental lines, because if, for example, authorization is by one person and his or her boss, there is greater opportunity for collusion, whereas if signoff from the internal audit staff is required, the possibility of collusion can be mostly, though not entirely, eliminated.

Rushby claims that "[i]ntegrity has received rather less attention than disclosure, although [disclosure] is arguably more important in some applications." While integrity is clearly a key aspect of both systems and the data that they handle, considerably more attention has been given over the past decade to disclosure in the form of unauthorized access to, and misuse of, sensitive data such as nonpublic personal information and intellectual property, as well as secret data of all types.

In the mid-1990s, when Rushby wrote his paper [9], distributed denial-of-service attacks were receiving less attention than integrity by researchers. A decade or so later, denial of service became a high-visibility, high impact form of attack, as in the 2007 attack on Estonia that took down government and banking websites, bringing their activities to a screeching halt [14]. The technology for such attacks, including the takeover of zombies and the expansion of botnets, have made denial of service more effective and have given the method greater coverage.

The Real-Time Systems Approach

Rushby [9] defines a *real-time system* as "one whose correctness depends not only on values of its outputs, but also on the times at which they are produced."

This approach concentrates on priority preferences of system tasks. These are important to detailed system design but are not specific to safety or security, as were the two prior approaches. In this regard, if a low-priority task is given preference over a higher-priority task, then the effect is similar to a denial-of-service attack.

Security-Critical and Safety-Critical Systems

The above definitions of security and safety originally from Boehm, represent a more useful and consistent view of the differences between security-critical and safety-critical systems than do the definitions of both Firesmith [7] and Barbacci [8] described above. This is because the latter definitions are somewhat restrictive. In general, safety-critical software systems can be compromised by intentional acts (not just random accidents) and security-critical systems can be compromised unintentionally (with no malice involved). The Firesmith and

Barbacci definitions of safety do not capture the essential concept that safety-critical systems should preferably *prevent* the harmful impact of both accidental and intended hazardous events rather than protect individuals from harm.

To illustrate this point, let us consider the analogy of some of the safety features in a modern automobile, where the antilock braking system (ABS) is designed to prevent or avoid potential accidents that might occur because of icy road conditions, whereas air bags are designed to protect drivers and passengers immediately after a collision has occurred. These two automobile systems are generally implemented separately, each using its own sensors to detect a dangerous condition and subsequently responding to hazardous situations independently. ABSs are intended to prevent accidents, whereas air bags are designed to protect driver and passengers when an accident does occur. In general, security measures are designed to prevent or avoid the consequences of an attack on a system, or its misuse, whereas safety is all about protecting assets (particularly, but not limited to, human beings) in the event that a breach or failure occurs.

Furthermore, Firesmith [7] considers survivability to be the combination of safety and security. It can be argued, however, that survivability has to do with effective responses following the inability of systems to withstand attacks or prevent dangerous situations resulting from system failure. Therefore, one can usually make a tradeoff between the costs of prevention (or protection) and recovery (or survivability) costs. As shown in Figure 4.2, the determination of these costs can provide the analyst with an optimal point at which the total costs are minimized.

In many situations, a greater investment in prevention may allow less to be spent on response and recovery as a means of attaining a desired level of dependability and integrity.

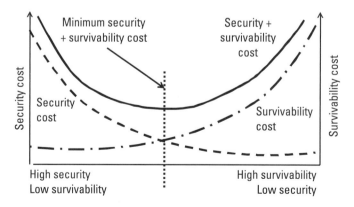

Figure 4.2 Protection versus survivability costs. (Source: "The Dynamics of Privacy Risk," *ISACA Journal,* Vol. 3, 2004 ISACA. All rights reserved. Used by permission.)

Summary and Conclusions

The goal of this chapter is to create a model of safety-critical and security-critical systems in the mind of the reader. While many researchers use the terms safety and security interchangeably, we choose to make a clear distinction between the two, with security measures being used to protect sensitive data and intellectual property within applications and databases, and safety measures being used to ensure that malfunctioning or failure of the system does not harm humans or the environment. While there are many parallels between the two in terms of concepts relating to attacks, defenses, and the like, there are major cultural and technical rifts among the engineers who design and operate each type of system.

The intention of presenting such clear-cut differences is to emphasize to those involved in each type of system that, especially, when security-critical and safety-critical systems are combined into systems of systems or cyber-physical systems, there is a need to share information and approaches and to collaborate across the whole security-safety spectrum. If this is not done, then the resulting systems will be deficient in their security and safety attributes—or even, in the worst cases, both sets of attributes.

Endnotes

[1] Anderson, R. J., *Security Engineering: A Guide to Building Dependable Distributed Systems, Second Edition,* Indianapolis, IN: John Wiley & Sons, 2008.

[2] On February 23, 2012, the Defense Advanced Research Projects Agency (DARPA), which is part of the United States Department of Defense, issued a Broad Agency Announcement (BAA) titled "High-Assurance Cyber Military Systems (HACMS)." In the BAA, *high assurance systems* are defined as those systems that are "functionally correct" and satisfy "appropriate safety and security properties." Unfortunately, the BAA does not define safety and security explicitly. The BAA is available at https://www.fbo.gov/index?s=opportuni ty&mode=form&id=925ab03b1bb59b1ac484e5a240b77097&tab=core&_cview=0, last accessed on July 12, 2012.

[3] Axelrod, C. W., "Trading Security and Safety Risks within SoS [Systems of Systems]," *INCOSE Insight,* Volume 14, No. 2, July 2011.

[4] Carter, A-L., *Safety-Critical versus Security-Critical Software,* 2010, available at http://www.bcs.org/upload/pdf/safety-v-security-report.pdf. Accessed July 12, 2012.

[5] There is a large body of material on the smart grid, not only in technical publications but also in discussions of cyber security by legislators. For example, the bill for "The Cybersecurity Act of 2012," which was introduced into the United States Senate on February 14, 2012, expresses deep concerns regarding the ability of the proposed Smart Grid to withstand cyber attacks. The bill is available by going to www.hsgac.senate.gov and searching for bill S. 2105. An overview of the Smart Grid is available on the U.S. Department of Energy website at http://energy.gov/oe/technology-development/smart-grid. Both websites were accessed on July 12, 2012.

[6] Barnes, J. G. P., "Ada," in *Avionics: Elements, Software and Functions,* C. R. Spitzer (ed.), Boca Raton, FL: CRC Press, 2007, pp. 15-1–15-50.

[7] Firesmith, D. G., *Common Concepts Underlying Safety, Security, and Survivability Engineering,* Software Engineering Institute, Carnegie Mellon University, Technical Note CMU/SEI-2003-TN-033, 2003. Available at http://www.sei.cmu.edu/library/abstracts/reports/03tn033.cfm. Accessed July 12, 2012.

[8] Barbacci, M. R., *Software Quality Attributes and Architecture Tradeoffs,* Software Engineering Institute, Carnegie Mellon University, 2003. Slide presentation and notes available at http://www.ieee.org.ar/downloads/Barbacci-05-notas1.pdf. Accessed on July 12, 2012.

[9] Rushby, J., *Critical System Properties: Survey and Taxonomy, SRI International, Technical Report CSL-93-01,* 1993. Available at http://www.csl.sri.com/users/rushby/papers/csl-93-1.pdf. Accessed July 12, 2012.

[10] Laprie, J. C., (ed.), *Dependability: Basic Concepts and Terminology,* New York: Springer-Verlag, 1991.

[11] Leveson, N. G., and P. R. Harvey. "Analyzing software safety," *IEEE Transactions on Software Engineering,* September 1983, Vol. SE-9, No. 5, 1983, pp. 569–579.

[12] See "Japanese worker inside stricken reactor recalls quake," by Terril Jones, March 26, 2011, at http://in.reuters.com/article/2011/03/25/idINIndia-55893620110325, last accessed on July 12, 2012.

[13] Clark, D. D., and D. R. Wilson. "A Comparison of Commercial and Military Computer Security Policies," *Proceedings of the Symposium on Security and Privacy,* Oakland, CA, IEEE Computer Society, April 1987, pp. 184–194. Available at http://groups.csail.mit.edu/ana/Publications/PubPDFs/A%20Comparison%20of%20Commercial%20and%20Military%20Computer%20Security%20Policies.pdf. Accessed July 12, 2012.

[14] See http://en.wikipedia.org/wiki/2007_cyberattacks_on_Estonia, last accessed on July 12, 2012.

5

Engineering Secure and Safe Systems, Part 2

I find it very frustrating when we are still talking about the same debugging techniques that were 'old' when I started teaching in 1971 [1]
—Marvin V. Zelkowitz, Fraunhofer Center for Experimental Software Engineering

Introduction

We have discussed previously how the practice of software engineering appears to be inadequate in the face of increasing security, safety, and other so-called nonfunctional requirements, such as dependability, performance, and resiliency. It is indeed disconcerting to read frequent articles by Zelkowitz [1] and Parnas [2, 3], who have been working in the field of software engineering for four or five decades, in which these pioneers in the field complain about the lack of progress that has been made in the software engineering space. The shortfall is particularly noticeable as it relates to safe and secure software systems. It is these latter systems that are the focus of this chapter.

The situation is even more disturbing when one considers that software-intensive systems are becoming more complicated and difficult to manage, especially with diverse systems increasingly being combined into systems of systems and cyber-physical systems. This software-system revolution is making much greater demands than before on engineers, as they are required to broaden the scope of their knowledge and professional efforts if they are to stand a chance of addressing the complexity and diversity of new systems. Furthermore, society is becoming so much more dependent on software systems, and as a result, system

73

failures have a much greater impact than previously on our personal lives and the health of the economy.

The gap certainly appears to be growing between the complexity of and dependency on software-intensive systems, and our ability to ensure that such systems are safe and secure enough to prevent humans from being harmed and to protect information assets from unauthorized access and from misuse. This situation does not bode well for a future where the gap continues to widen, perhaps exponentially. Until and unless software systems professionals in particular and society in general make enormous progress towards narrowing that gap and eventually eliminating it, we will not see sufficient improvements to stave off catastrophic failures.

One of the most dramatic examples to date of the failure of a computerized information system occurred on June 19, 2012, when the Ulster Bank's automated processing system, which usually handles some 20 million transactions per day, suffered a catastrophic failure. The outage prevented access to and destroyed the integrity of the bank accounts of more than 100,000 customers. A month later, many of the problems that resulted from the initial system failure remained unresolved. The ultimate direct cost of this incident will likely be in the tens of millions of Euros (if not greater), if the consequential costs of loss of trust in the dependability of the computer systems and the impact on the banks reputation are included [4].

The main goal of this book is to raise awareness of the dangers that we are currently experiencing and that lie ahead. These dangers may result from inadequate software systems engineering, particularly as it relates to safety and security. Such threats will have much greater impact if we are not both willing and able to make rapid and significant improvements in the processes and procedures used for building security, safety, and other nonfunctional attributes (such as resiliency) into software-intensive systems. For the effort to be effective, we need to see major increases in the number of subject-matter experts who can take over the responsibility of ensuring the safety and security of critical software-intensive systems.

Another goal of this book is to guide those who recognize the vastness and importance of the problem towards meaningful ways of achieving greater assurance that systems will not cause physical harm (safety) and will protect the information that we most value (security).

In this chapter, we trace the history of various branches of engineering; this history puts today's software crisis in perspective, serves to explain why we find ourselves in the dilemma of recognizing what it will take to establish acceptable levels of security, and having to accept, albeit very reluctantly, the limits of safety and our ability to get decision-makers to make available the required resources to achieve such levels.

Approach

If we use the analogy of a victim of a medical emergency, it is first necessary to stabilize the patient before administering curative treatments. So it is with security-critical and safety-critical software-intensive systems. Before we can expect to prevail over the attackers, we must first prevent the gap between attackers and defenders from widening further; we can then work on reducing and eventually eliminating the gap. Perhaps we might even get ahead of the attackers. Continuing to do as we are now doing will not allow us to escape or reduce the impact of attacks.

Figure 5.1, which is not to scale, illustrates the exponential rise in security and safety requirements and system complexity as systems are combined, particularly cyber and physical systems. The figure also shows the relatively slow increase in software assurance efforts and capabilities, indicating that the gap is widening at an accelerating pace. As some assert with global warming, even if there were an immediate resolution, the problem will continue to grow rapidly due to the enormous momentum that is already in place. The best one can hope for under such circumstances is to slow the growth in the divergence between the increases in adverse effects and the ability to prevent them in the hope that new solutions will appear that are effective in reducing, and then eliminating, the gap.

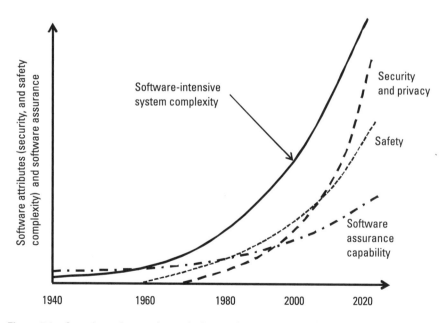

Figure 5.1 Security, safety, and complexity requirements versus software assurance capabilities. (From Axelrod, C. W., "The Dynamics of Privacy Risk," *ISACA Journal*, Vol. 3, © 2004 ISACA. All rights reserved. Used by permission.)

The approaches recommended here are less about protection and more . concerned with prevention. Our main focus is on avoiding safety and security issues in the first place rather than trying to protect against them. After all, if software-system engineers were able to and were allowed to design, develop, and implement high-assurance software-intensive systems from the outset, many—if not virtually all—subsequent mitigation efforts could be avoided, as could the need to respond reactively. It is indicative of the extent of the problem to see requests for proposal from the Defense Advanced Research Projects Agency (DARPA) for new approaches for ensuring that software-intensive systems exhibit appropriate safety and security properties. For example, DARPA-BAA-12-21 on the subject "High-Assurance Cyber Military Systems (HAC-MS)," is looking for "clean-slate approaches" to assuring the safety and security of embedded systems control vehicular operations [5].

While we have the technical ability to build safe software-intensive systems, as has been demonstrated in the avionics and military field, there are many other situations in which the heavy-duty methods that are used to ensure that systems will not fail and are invulnerable to cyber attacks cannot be justified. However, it is likely that cost-benefit analyses will not show a positive return on investment for expenditures on such intense risk mitigation activities. Whereas such results might be due to an underassessing the full impact of losses, which might be incurred were the system to be attacked, malfunction, or fail, it is tough to convince decision-makers of the risks of something really bad happening until it actually happens, as described above in the case of the malfunctioning of the Ulster Bank's automated processing systems.

Despite the high frequency of system failures and the consequent costs, we might be on the wrong track if we expect material changes to be made for all but the most critical of software systems. Often what is merely "good enough" is considered acceptable to decision-makers. If that is the case, then attempts to achieve improvement beyond a mediocre level of assurance will not win them over.

Reducing the Safety-Security Deficit

The unrelenting increase in the demand for more functional capabilities from software products and software-intensive systems, together with the aggregation and combining of systems, are leading to systems of systems and cyber-physical systems of ever greater complexity and with higher safety and security needs. Yet we continue to fall behind in our ability to withstand attacks and restrict damaging side effects, as evidenced by the growth in reports of attacks and software system failures. Experience has shown that the juggernaut of systems development and implementation will not stand by and wait for software engineers, in particular, to get their act together. It hasn't happened before—it is unlikely that it will happen in the future unless tremendous pressure is brought to bear.

In many ways, the states of software system security, and to a lesser extent software system safety, are analogous to the national debt situations that we see in many developed countries. The spending on security and safety might be considered "deficit spending" as the demands for security and safety increase, and yet the total security and safety debt inevitably increases as the impact of attacks and software failures increases over time. The security and safety debt continues to grow as security and safety deficits are incurred on a regular basis. Not until we see ongoing annual safety-security surpluses (meaning that the net effect of expenditures on security and safety exceeds the losses to individuals, organizations, and societies), can we hope to see the total safety-security debt reduced. As much as politicians and society favor deferring the need to reduce the financial debt until some distant future time (when somebody else will have to foot the bill), it would seem that security and safety engineers and those responsible for making security and safety decisions are willing to defer addressing existing security and safety deficits at the present time. Of course, this will only lead to much greater costs in the future as we pay "interest" at a very high rate on the security-safety debt each time that an attack is successful or a malfunction or failure occurs. This suggests that even the exponential increases illustrated in Figure 5.1 might actually be conservative.

Game-Changing and Clean-Slate Approaches

In many situations, there is a need to develop software assurance methods to enable the state of the art to accelerate at a rate that is much faster than the speed of functional innovation. That is to say, the rise in software complexity and criticality currently far exceeds our ability to test the software effectively, especially when diverse components are bound together to form systems of systems or cyber-physical systems.

The common response to this problem is to look for opportunities to build security and safety measures into systems throughout their development life cycles, especially during the earlier phases. While such an approach has a good chance of improving the quality of future systems in terms of security and safety, it does little or nothing to enhance existing systems. This is because of the much greater effort required for testing so-called legacy software and the orders of magnitude higher costs of retrofitting fixes (or patches) and reprogramming and retesting in-place production systems.

Consequently, many practitioners are looking to researchers to come up with game-changing, clean-slate approaches to produce high-assurance systems that are not only functionally correct, but are also both safe and secure. Many of the suggested game-changers, such as rapidly changing the security environment in order to get a step ahead of attackers, providing economic incentives to engender good security, and creating a flexible distributed trust environment, are worthy goals. These approaches or themes were suggested at a May 19, 2010 conference hosted by the Networking and Information Technology Research and Development (NITRD) Program and are summarized in a presentation with the title "Toward a Federal Cybersecurity Research Agenda: Three Game-changing Themes," which was given at the conference [6].

Despite the appeal of creating such "silver bullets," realistic, operable, and cost-effective approaches for meeting these and similar goals do not appear to have been developed, at least in the public domain, and it is highly unlikely that such all-encompassing solutions will be created in the near future, given the current state of software systems development and implementation. However, were such a miracle to happen, the creation of innovative technologies, which are also cost-effective across a broad range of public-sector and private-sector systems, would certainly revolutionize the way in which software safety and security attributes are handled and could quell many of the concerns that are now being voiced.

Appeals to researchers to increase the safety and security of cyber systems and to improve the verification process are not limited to the NITRD case mentioned above. Requests for proposals have also come from the Information Innovation Office (IIO) of DARPA and others. For example, in the first quarter of 2012, DARPA issued two Broad Agency Announcements (BAAs) of particular interest. One is DARPA-BAA-12-21 [7], which has a goal of receiving proposals for creating "technology for the construction of high-assurance, cyber-physical systems, where high assurance is defined to mean functionally correct and satisfying appropriate safety and security properties." The requestors are looking for original, innovative technologies, not ones that have been previously developed or are merely extensions of prior work.

Another BAA of interest is DARPA-BAA-12-17 [8], which is about crowd-sourcing the software verification process. In this case, DARPA is looking for "innovative approaches that enable revolutionary advances in science, devices, or systems." Here again, evolutionary improvements are excluded.

It would appear that DARPA is attempting to circumvent current software assurance technologies and practices, which they seem to question with respect to the methods' ability to ensure that software is safe and secure. While it is disconcerting to realize that the motivation behind such DARPA requests is that current approaches are inadequate, it is reassuring that the inadequacies have been recognized and that there is a willingness to fund reesearch into new technologies. It seems that we will have little choice but to rely upon the resulting technologies to get ahead of the security and safety deficiencies in past, present, and future mission-critical, software-intensive systems.

If researchers do not produce ground-breaking, cost-effective solutions—and there is no guarantee that they will—then it will be necessary to revert back to older safety and security methods and try to enhance them. If that becomes the case, then we will have to put forth a huge effort and devote enormous amounts of resources into building security, safety, dependability, integrity, resiliency, and other attributes into future software-intensive systems, starting from the earliest phases of the software systems development life cycle. This approach will certainly involve enormous expense and effort. There is also a need to develop a cadre of subject-matter experts able to accomplish the enhanced requirements. Retrofitting security, safety, and the like into existing systems would incur an even greater effort and expense because of the need for software engineers to be familiar with obsolete computer languages and operating systems, in addition to the capabilities required for up-to-date methods and technologies.

The combination of retrofitting older systems and building security, safety, resiliency, and the like into new systems would cost so much that such an effort is extremely unlikely to be justified and subsequently funded. Rather, a more acceptable approach would be to replace existing systems over time with new systems of greatly improved quality with respect to security, safety, and the like, assuming that there is the time and determination to even get this done. Even a lesser initiative would have huge difficulties in finding sources willing to fund the effort.

Only some overriding threat—such as existed for the Year 2000 or Y2K mitigation effort, when there was major concern about the risk of system malfunctions and failures due to systems misrepresenting the date over the century rollover—would motivate public and private sectors to fund remediating currently-operating and in-development software-intensive systems. Y2K remediation was estimated to have cost in the $200 to $300 billion range, although much of those expenditures consisted of accelerating the replacement of existing systems with new systems. Because the Y2K fixes could be made comparatively easily without having to understand much about the functionality of the software being corrected, that effort can be expected to have been orders-of-magnitude less costly than the effort required for implementing a broad array of security and safety requirements to existing and new systems. It is likely that bolting security and safety on to all existing mission-critical systems could cost in the *trillions* of dollars. Doing so for future systems could add billions of dollars to projects, although such costs could be offset against reduced vulnerability and fewer successful attacks. It is difficult to envisage any potential catastrophe that could precipitate such expenditures—until it actually happens.

Furthermore, it is unlikely that even building security, safety, and resiliency into new systems will be universally accepted, particularly by private industry. Some companies, such as the Microsoft Corporation, have notably embraced the concept of trustworthy computing, effectively replacing prior products with more secure new products. However, only a few software manufacturers appear to have followed Microsoft's lead. Some 19 of 51 companies that are participated in the fourth iteration of the Build Security in Maturity Model (BSIMMH) initiative were independent software vendors (ISVs). Such firms as Adobe, Intuit, Microsoft, SAP, and VMware are included in the BSIMM community of companies. Participation in the BSIMM projects is considered indicative that these firms are at the forefront of security practices [9]. However, the goal of universal trustworthy computing is unlikely to be realized unless there is a push from government in the form of financial incentives, threats of legal and regulatory action, or both.

On February 14, 2012, a cyber security bill, S.2105, was introduced into the U.S. Senate by Senators Lieberman, Collins, Rockefeller, and Feinstein. The bill was titled "Cybersecurity Act of 2012," and it attempted to address concerns about cyber attacks on industrial control systems, particularly those affecting the electrical energy industry and other utilities [10]. Such safety-critical systems are increasingly being connected to public telecommunications networks, such as the Internet. The Cybersecurity Act of 2012 aimed to address issues arising from cyber threats and common vulnerabilities in software systems. As would have been expected, especially in the then-current hostile political environment, the bill almost immediately ran up against opposition because of the restrictions that it would impose and the costs that would be incurred. If the history of the recent attempts to introduce Federal cyber security legislation is any guide, it is unlikely that this bill will find its way into law in its current form, and any compromise bill is likely to be dragged out over time and considerably watered down.

An earnest push for action on national cybersecurity legislation will likely not occur until the situation has become bad enough to cause public outcry and political response. Vice Admiral Michael McConnell (USN, Ret.) testified on this at a hearing of the U.S. Senate Committee on Commerce, Science, and Transportation on February 23, 2010, saying that he believes that no serious action will be taken on behalf of cyber security until and unless a catastrophic cyber event were to occur. The Committee was chaired by Senator John D. (Jay) Rockefeller IV, and was on the topic "Cybersecurity: Next Steps to Protect Our Critical Infrastructure" [11].

Meanwhile, apparently all we can do is just try to prepare for the day when such an effort will be endorsed out of necessity, as it will when a catastrophic event or series of events takes place. This is by no means a desirable way of addressing the issue, but as of now, there seems to be little choice. By way of preparation, we need to sort out the details of how, in an ideal world, safe and secure software-intensive systems should be designed, developed, implemented, run, and discarded. In that way, there will be some basis for quickly invoking the better methods needed to deal with the problem at the time. Such an approach is far from ideal but, because there is little chance of establishing appropriate standards and funding the implementation of those standards without a major driver, it will have to serve the purpose, even though it will cost many times what it would cost if the mitigation efforts had been done in advance.

A Note on Protection

The general approach to security is to defend systems against attacks, thereby protecting information assets. With respect to safety, not only do we want systems not to harm humans, other life forms, and the environment, but we also look to protect potential victims directly through secondary means, such as manual overrides in the event that systems malfunction or fail due to intentional or accidental incidents.

Most current efforts are geared to prevention. Typically, we install intrusion detection and prevention systems to protect software systems against malicious attacks. Antivirus software is used to detect and intercept malware before it can insert itself into legitimate computer systems. Safety-oriented systems are prevented from doing damage by using a specific set of failsafe methods to protect individuals, groups, and the environment.

Not nearly enough money and effort is being applied to deterrence and avoidance, relative to funds allocated to protective measures. However, avoidance and deterrence are usually far cheaper to implement than preventative approaches. For example, if the British royal crown jewels were to be secreted away in an unknown location, far less effort and expense would be needed to guard and protect them than if the real crown jewels were on display. However, it can be argued that displaying them is worth more than what it costs to protect them, which is why they are put on view in all their glory. The main argument in this case against avoidance by not exhibiting the jewels is that the public has a right to see them and is willing to pay to do so. Nevertheless, facsimiles could be substituted, thereby avoiding much of the security issue, although the experience of viewers would be diminished (if they were, in fact, aware that they were looking at fakes), except if the viewing public believed them to be real because of the extraordinary security precautions being taken.

Deterrence and avoidance, in particular, are the keys to attaining levels of security and safety that professionals in the field would like to see. However, since protection is seen as a responsibility of safety and security professionals, and deterrence and avoidance relates more to business decisions, the latter are seldom included in the systems engineers' toolbox. Often software-intensive systems are breached, or fail in a harmful way, not because they were not protected against unwanted actions, but because they are being used for purposes not originally envisaged, approved, or allowed for.

A prime example of this latter condition is the Internet itself. Originally called ARPAnet, the Internet was developed by DARPA as a highly resilient network that could be relied upon to continue to operate even if substantial segments of the network were disabled or destroyed. The Internet was not designed with security in mind, nor safety for that matter, because the original users, mostly academics, knew and trusted one another and because if users failed to honor appropriate security requirements, they were banned from using the network. Ironically, many software system compromises usually occur because of unauthorized or unexpected access, use, and operations that were not envisaged by the creators of the software.

Deterrence and avoidance must be included as means of reducing software system risk if we are to have any hope of meeting the challenge. Such measures are often unpopular because they frequently mean restricting or removing prior privileges and uses. Also, those enforcing the restrictions will likely be accused of stymieing innovation. Yet deterrence and avoidance, if thoughtfully, carefully, and properly applied, can be much more cost-effective than protective measures. At some point, when it is clear that protection is insufficient and that critical infrastructure systems are on the verge of collapse, society might be willing to take the difficult steps of restricting functionality and use of such critical systems and eliminating ready access to sensitive information. Until then, we are left with trying to improve the quality of current protective measures and work on encouraging them to be used and enforced.

Despite the expected lack of enforcement and adherence, we include deterrence and avoidance measures when considering a holistic approach to achieving safety and security goals. It is expected that, at some future time, society may well begin to focus on deterrence and, in particular, on avoidance as disappointment with protective measures increases.

It should be noted that deterrence and avoidance of inappropriate access to, misuse of, and damage to software functionality and data can often be built into software design and development. In particular, the design of a software-intensive system needs to incorporate the ability to restrict access to specific functions and data. For example, while a third-party identity and access management system might front-end a particular application, the application software itself needs to interpret user access restrictions properly and have the ability to block inappropriate functionality and data access. In other cases, deterrence might be introduced, for example, through warning messages letting users know that their aberrant behavior has been identified and recorded, and that action will be taken in the event that users exhibit such behavior again. These techniques are often less difficult and expensive to implement than are methods for protecting data, such as encryption.

Safety-Security Governance Structure and Risk Management

We also address the importance of having an effective governance structure for the design, development, and implementation of software systems. In addition to the usual recommended project management processes, which are common in the literature, it is critical to ensure that all the necessary subject-matter experts are brought in at the appropriate stages of the software systems development life cycle (SSDLC).

We also emphasize the need to establish a risk management individual or group that will ensure as much as possible that both functional and nonfunctional requirements are given their appropriate weight when making tradeoffs among project cost, quality, and timeliness.

Not only are software-intensive systems becoming more complex and interconnected, but many software systems are transitioning into "the cloud." These trends increase the need for both coordination and collaboration among systems engineers, software engineers, safety engineers, and security professionals throughout the entire development life cycle. Unfortunately, many of these groups do not currently communicate with one another; as a result, software-intensive systems often lack the quality that a collaborative effort might well have produced.

The governance and risk management structures and practices described in this book are designed to foster such communication and collaboration. The proposed approach will also ensure that all significant factors for optimizing so-called nonfunctional requirements of software systems such as security, safety, resiliency, survivability, and sustainability are at least considered. In the best of circumstances, these attributes will be incorporated into the processes used to create software systems, and will thereby protect internal assets, prevent the system from damaging others, remain operational during destabilizing and damaging incidents, and ensure the integrity of sensitive information and systems. The approach followed here is first to consider life cycle structure, then processes and finally the roles and responsibilities of those who need to be involved.

An Illustration

In mid-January, 2012, passengers in a British Airways flight from Miami to London heard an announcement that the airplane was about to crash into the ocean. According to a July 1, 1999 article "BA False Alarm Strikes Again," the issuance of the prerecorded alarm about a British Airways aircraft ditching into the sea is not new. In fact, such incidents were reported to have occurred three times within four months in 1999, according to the article [12]. The same thing happened again on a flight from London to Hong Kong, as mentioned in the January 17, 2012 article "Oops: Plane Mistakenly Warns Passengers of

Crash" about the January 2012 incident [13]. In another January 17, 2012 article "British Scareways" by Robert McAuley in *The Sun* newspaper, the prior incident occurred in August 2010 [14]. In some cases, the malfunction was attributed to a computer glitch; in others cases, human error was given as the cause. One would have thought that, had a proper root-cause analysis been conducted, the reasons for such occurrences would have been determined and the systems modified to prevent subsequent repetition of the false alarm.

Fortunately, the announcement system had not been triggered by an actual physical emergency or a failure of the aircraft's control systems. However, it does point to a failure in the coordination among the aircraft's information systems, control systems, and their human operators, particularly since similar incidents have occurred at least four times previously on British Airways flights.

The General Development Life Cycle

We shall be focusing here on software-intensive systems rather than just application software. That is to say, we are including system software, firmware, hardware, facilities, and human elements in the mix.

However, before we get into the details of safe and secure software system engineering, we will look at the origins of various engineering disciplines and how these disciplines evolved; the purpose being to illustrate how certain gaps might have developed over time.

Even today, traditional systems engineering is more commonly applied to the construction of facilities (office buildings, data centers, etc.) and the manufacturing of equipment than it is to software development. Nevertheless, the use of project management tools for software development has a fairly long history, stretching back to the 1940s, with wider adoption in the business world (as opposed to the military) beginning in the 1960s as those with military and government backgrounds entered the private sector.

The earlier and more extensive use of project management in construction and manufacturing may be because these fields are more closely aligned with longer-standing engineering practices, such as civil, mechanical, and electrical engineering, which have all been established as professional disciplines for more than a century, as indicated in Table 5.1. In his book, Gawande [15] provides an interesting description of the sophisticated project management tools used by construction companies to coordinate the many activities and capture resource use for complex projects, such as the construction of a skyscraper.

Software engineering is a relatively new discipline with a history spanning only six decades. Software safety engineering has a somewhat longer history of perhaps 40 to 50 years as compared to software security engineering, which has only been active for some 20 years or so. This latter difference is likely because

Table 5.1

Approximate Timeline for Emergence of Formal Engineering Disciplines

Discipline	Date of Origin	Description of Events Related to Origin	Main Sources
Civil engineering	1760s	Englishman John Smeaton was the first formal civil engineer.	See H2G2 website at http://h2g2.com/dna/h2g2/A918371
Mechanical engineering	1847	Institution of Mechanical Engineers (IMechE) was formed in Birmingham, England, by George Stephenson.	See IMechE website at http://heritage.imeche.org/Home
Electrical engineering	1884	Formation of the American Institute of Electrical Engineers.	See IEEE Long Island Chapter Pulse Newsletter, page 8 at http://www.ieee.li/pulse/pulse_2012_02.pdf
Safety engineering	1911	Formation of The American Society of Safety Engineers (ASSE) in 1911.	For history of safety engineering, see ASSE website at http://www.asse.org/about/history.php
Systems engineering	Early 1940s	The term systems engineering dates back to Bell Telephone Laboratories in the early 1940s, The concepts of systems engineering within Bell Labs back to early 1900s. There were major applications of systems engineering during World War II. The first attempt to teach systems engineering, as we know it today, came in 1950 at the Massachusetts Institute of Technology (MIT) The course was taught by Mr. Gilman, Director of Systems Engineering at Bell Telephone Laboratories.	See INCOSE website at http://www.incose.org/mediarelations/briefhistory.aspx Quotation from D. Buede, *The Engineering Design of Systems: Models and Methods*, Wiley, 2000.
Software engineering	1950s	Pioneering era 1955-1965.	Review of "An early history of software engineering" by Robert L. Glass available at http://www.cs.colorado.edu/~kena/classes/5828/s99/comments/srinivasan/01-29-1999.html
System safety engineering	1960s	First application in military (1962) Air Force MIL-STD-882 in 1969.	For history of system safety in the U.S. military, see http://www.dtic.mil/ndia/200Ssystems/wednesday/mcallister.pdf
Security engineering	1970s	Issue of computer security first arose as individuals began to break into telephone systems. Security (or, more specifically, cryptography) can be traced all the way back to Roman times when Julius Caesar enciphered his dispatches.	See http://ecommerce.hcstip.info/pages/249/Computer-Security-HISTORY-COMPUTER-SECURITY-PROBLEMS.html
Software systems safety engineering	1970s	Introduced into U.S. space program. Avionics pioneered in 1970s.	
Software systems security engineering	1990s	Early pioneers include such thought leaders as Gary McGraw and Ken van Wyk.	See the "Build Security In" website at https://buildsecurityin.us-cert.gov/bsi/home.html

Note: The website links contained in this table were all accessed on July 15, 2012.

software developed for aircraft and spacecraft required that the negative impact of system failure on human beings be considered, whereas interest in the security of software only began to receive significant attention with the advent of distributed systems and the Internet. There was some interest in security prior to the arrival of distributed systems and the Internet, but it was mostly limited to restricting logical access to data stored on mainframes using off-the-shelf software packages such as RACF, ACF2, and TopSecret. There were also a limited number of such access management systems for minicomputers. At that time, logical access to computer systems was through "dumb terminals," which were connected via dedicated cables to computer systems. Physical security methods were used to restrict access to rooms containing terminals and computer systems.

Structure of the Software Systems Development Life Cycle

The Department of Defense document on systems engineering [16], though more than two decades old, may still be considered to be the definitive text on the topic, because the Defense Acquisition University (DAU) website currently lists the document and it does not appear to have been superseded. However, more recent texts on the subject have been issued by NASA [17] and Haskins [18].

Haskins [18] provides useful comparisons among a half-dozen different life-cycle models. The comparisons are summarized at a higher level in Table 5.2.

As Table 5.2 shows, the various *systems engineering* life cycles follow roughly the same pattern. However, some life cycles emphasize particular phases over others. For example, the DoE cycle, as described in [19] does not include an explicit deactivation phase, but has more extensive planning phases than do other life cycles.

We now look briefly at the structure and phasing of software engineering life cycles. They tend to follow the phases shown in the second column of Table 5.2. However, most researchers emphasize life cycles of internally developed software because there is much less ability to control third parties that produce and sell software. On the other hand, system integrators, as referenced by Haskins in Table 5.2, also go through similar life cycles as do organizations that develop software systems from scratch. In Table 5.3, we show structures for the software engineering life cycle, much as systems-engineering life cycle structures were shown in Table 5.2.

An immediately apparent difference between the systems and software engineering life cycles (comparing Tables 5.2 and 5.3) is that the latter does not show a retirement or decommissioning phase. While some may consider this

Table 5.2
Comparison of Systems Engineering Life Cycle Phases by Source

ISO 15288:2008*	NASA Systems Engineering Handbook [17]	High-Tech Commercial System Integrator [18]	High-Tech Commercial Manufacturer [18]	Department of Defense (DoD) 5000-2 [20]	Department of Energy (DoE) [19]
Concept	Formulation: -Concept studies -Concept & technology development Approval: -Preliminary design & technology completion	Study: -Requirements definition -Concept definition -System specification -Acquisition preparation	Study: -Requirements definition -Product definition -Product development	User needs Presystems acquisition: -Material solution analysis -Tech develop	Planning: -Preplanning -Preconceptual planning -Conceptual design
Development	Approval: -Final design & fabrication	Implementation: -Source selection -Development	Implementation: -Engineering model -Internal test	Systems acquisition: -Engineering/manufacturing development	Execution: -Preliminary design -Final design
Production	Implementation: -System assembly, integration & test -Launch	Implementation: -Verification	Implementation: -External test	Systems acquisition: -Production & deployment	Execution: -Construction
Utilization Support	Implementation: -Operations and sustainment	Operations: -Deployment -Operations -Maintenance	Operations: -Full-scale production -Manufacturing, sales, support	Sustainment: -Operations & support	Mission: -Acceptance
Retirement	Implementation: -Closeout	Operations: -Deactivation	Operations: -Deactivation	Sustainment: -Disposal	Mission: -Operations

*The ISO 15288:2008 standards focus on processes rather than framework or structure. The example of a life cycle shown in the table is provided in the standards document.

Table 5.3
Comparison of Software Engineering Life Cycle Phases by Source

Waterfall Model (Royce)*	Spiral Model (Boehm)**	Braude and Bernstein [21]	Merkow and Raghavan [22]	McGraw [22]	Christensen and Thayer [24]	ISO/IEC/IEEE 12207-2008 [25]	Ferrell and Ferrell [26]
—	—	—	—	—	Recognition of need	—	—
—	—	Planning	—	—	Acquisition decision and strategy	Acquisition supply	Planning process
Requirements	—	Requirements analysis	Requirements	Requirements and use cases	Specifications	—	Requirements
Design	Product design	Design	Design	Architecture and design	Design	—	Design
—	Development plan	—	—	—	—	—	—
—	Risk analysis	—	—	—	—	—	—
—	Prototypes	—	—	—	—	—	—
—	Final design	—	—	Test Plans	—	—	—
Code & Unit Test	Code	Implementation	Development	Code	Implementation	Development	Code
Test & Integration	Test	Testing	Test	Test and test results	Acceptance	—	Integration Verification Configuration management Software quality assurance Certification liaison
—	Ship	—	—	—	Release to field	—	—
Operation & Maintenance	—	Maintenance	Deployment	Feedback from the field	Maintenance	Operation	—
—	—	—	—	—	Requirements [sic.]	Maintenance	—

*See, for example, Braude & Bernstein [21], pages 37-38.
**See, for example, Braude & Bernstein [21], pages 44-46.

unimportant, it illustrates a difference in philosophy and represents a major gap in the creation and disposal of software. It might be explainable from the physical difference between systems in general, and software in particular, where the physical realization of software consists of data bits on some storage medium that can often be deleted with a single software command, whereas equipment disposal is often much more involved.

Another difference that is not as well illustrated by comparing the structure tables (but which will become more apparent when comparing processes) is the application of verification and validation (V&V) rather than testing. V&V is usually a much more intensive process than general software testing and generally a requirement for military, avionics, and similar systems. In fact, independent V&V (or IV&V), which is performed by an independent third party, is frequently mandated for such systems. As mentioned previously, Braude and Bernstein [21] define the terms as follows:

- *Verification*: Ensuring that each artifact is built in accordance with its specifications.

- *Validation*: Checking that each completed artifact satisfies its specifications

That is to say, we verify that we are building the product correctly, mostly by means of inspections and reviews; we validate that we are building the right product, mostly by testing [27].

Life Cycle Processes

Importantly, in Christensen and Thayer [24], the authors distinguish between systems engineering and project management as follows:

- *Systems engineering*: a capability within an organization that might be identifiable as a separate entity performed by a combination of individuals, such as system architects, process architects, or lead designers. It is applied to technical activities as they relate to the development and sustainment of systems.

- *Project management*: the function (or role) of project manager exists for any project. The individual is responsible for managing the activities such as process planning and control.

A basic attribute, or function, of systems engineering is the formal management of specifically-structured systems development projects. Such discipline provides a substantial measure of assurance that customers and end users (if they are different from customers) will ultimately receive completed systems that meet their specific requirements and that have been tested to ensure that they work as required and specified.

One might reasonably assume that customers and/or end users are able to determine whether or not a particular system satisfies their functional requirements (e.g., a bridge spans from point A to point B and has the capacity to handle given traffic volumes). However, typical customers (in the case of a bridge, the customer might be a municipality representing those looking to use the bridge) cannot be expected to have the expertise needed to ensure that the system meets nonfunctional requirements (e.g., for a bridge, such requirements might include the maximum weight of traffic and the stability of the bridge in response to crosswinds of different velocities). For software-intensive systems, it is similarly not to be expected that customers or end users can be relied upon to have specified appropriate safety and security requirements with acceptable completeness. When someone purchases an automobile, for example, he or she will look at factors such as space, comfort, fuel consumption, braking distance, and the like, in making a purchase decision; however, safety features are assumed to be given, although consumer rating organizations will report safety issues, including results of the National Highway Traffic Safety Administration (NHTSA) impact tests. Similarly, customers and end users of software-intensive systems cannot be expected to define safety requirements. Furthermore, when it comes to the security of on-board vehicle control systems, there are not yet standards against which to measure the security level, official tests are not conducted, and the purchaser is left to his or her own devices when it comes to determining and assuming the risk that the systems will be compromised by attackers.

For software-intensive systems, in general, the functional needs of customers and end-users do not usually include safety and security functional and nonfunctional requirements, and customers and end-users are not usually qualified to test systems for safety and security deficiencies. Some, such as Brenner [28], have suggested setting up an independent testing organization (equivalent to Consumers Union) to test and rate commercially available software. In other cases, specific industries, such as the U.S. financial services sector, have looked into setting up an industry software assurance laboratory. Among many issues with establishing such a testing capability are agreeing to who would fund the effort, which software systems would be selected for testing, which set of criteria would be used for testing, and how the results would be distributed.

It is therefore important to include subject-matter experts who are knowledgeable in both functional and nonfunctional areas on the systems development project team. However, in many cases, particularly for secure software manufacture, the appropriate experts are often not called in during the life cycle and are forced to try to bolt security and safety measures after the software system has already gone through development and testing. At this point there has to be considerable compromise, particularly as the costs of such belated efforts are many times what they would have been if the same security and safety features had been built in during earlier phases. In general, the problem is less for safe software systems than for secure software systems because the former are more often governed by specific standards and certification requirements.

At this point, we will take a more detailed look at systems engineering processes and related responsibilities of those managing and working on the systems. We draw heavily on guidelines developed by the space and military agencies as they were prominent in developing the field.

In Christensen and Thayer [24], the functions of systems engineering are defined to include the following:

- Problem definition;
- Solution analysis;
- Process planning;
- Process control;
- Product evaluation.

According to Parnas [3], the systems engineering process comprises the following functions:

- Requirements analysis;
- Functional analysis and allocation;
- Design synthesis;
- Verification;
- Systems engineering process outputs.

From the definitions of verification and validation by Braude and Bernstein [21], mentioned earlier, we see in the above process by Parnas that there is a phase in which systems are checked to see that they were built correctly (i.e., verification), but not that the systems necessarily address their specifications. This can result in very well-built systems that do not do what they were intended to do, and what is even more problematic, they might do what they are not supposed to do.

As indicated above, very similar structures and phasing have been applied to software engineering, as well as to software safety and security engineering. The main differences relate to the processes within each phase and the degree of emphasis placed on each. Just because a phase is recognized as part of the life cycle, it does not mean that it will be addressed adequately unless the requirements and the processes are clearly and completely specified from the start.

There are generally substantial differences in the application of particular effort and expertise as we move from general systems engineering to more specialized software engineering and then on to the engineering of safe and secure software. Deficiencies in the specific areas of software security and safety result more from processes that are inadequately adhered to, rather than inappropriate life cycle structures. Key quality areas are often given insufficient attention and resources, or are ignored completely, suggesting that additional subject-matter experts, skilled in areas such as application security and system safety may have been excluded from critical stages in the life cycle. Their skills and experience need to be brought to bear during critical phases of the life cycle if deficiencies in these areas are to be addressed and resolved.

Governance Structure for Systems Engineering Projects

Researchers are usually quite specific about what should be done, but often pay insufficient attention to how success in each phase might be achieved. Here the devil is in the details, and many of the shortcomings that we see in the safety and security of systems are due to researchers either not including or not even being aware of the range of skills and backgrounds needed to create a successful safety-critical or security-critical software system. Again, this disparity is becoming particularly common as complex systems of systems and cyber-physical systems comprising both safety-critical and security-critical components are evolving.

In Table 5.4, we show the differences among the disciplines of general software engineering, software safety engineering, and software security engineering. As can be seen from the table, it took several decades from when software engineering was created as a separate, independent area to formalize the inclusion of software safety, and several more decades for software security engineering to become recognized as its own domain. The relative maturity of each of these disciplines is reflected in the status of formal standards, establishment of professional associations, availability of expertise, and so on. The maturity is also reflected in the respective quality of software systems being produced, with software security engineering being somewhat weak in certain areas, such as the establishment of generally-accepted stringent standards, relative to software safety engineering.

Table 5.4
Comparison of Characteristics of Software Safety and Security Engineering

Characteristics	Software Engineering	Software Safety Engineering	Software Security Engineering
Origin	1940s	1960s	1970s
Scope	All software, plus firmware	Usually military, aviation, vehicle and industrial control systems	Critical public and private sector information systems
Professional associations	IEEE Computer Society (formed in 1946) www.computer.org Association for Computing Machinery (ACM; founded in 1947), www.acm.org	The International System Safety Society (founded in 1963). http://system-safety.org	Information Systems Security Association (ISSA), www.issa.org ISC2 (formed in 1988), www.isc2.org Computer Security Institute (CSI, founded 1974), http://gocsi.com SANS Institute (founded 1989), www.sans.org
Level of application	Moderate	High	Low
Existence of formal standards	Some, but not widespread particularly in the private sector	Many standards, but not consistent across public and private sectors	Few, mostly aimed at particular industries
Examples of standards	See, for example, *NIST SP500-204 High Integrity Software and Standards*, at http://hissa.nist.gov/pubs/nist204.pdf	DO-178B, *Software Considerations in Airborne Systems and Equipment Certification* (see page 77, http://www.rtca.org/downloads/ListofAvailableDocs-Sept2012.pdf, accessed October 7, 2012)	Payment Card Industry Data Security Standard (PCI DS3) ISO/IEC 27001 and 27002 are available from http://webstore.ansi.org
Education and training programs	Numerous college courses and certifications in software engineering are offered.	Usually included in engineering programs rather than in software training	Generally ad hoc training. A few college programs in software engineering are beginning to include security
Adequacy of available expertise	Substantial number of trained software engineers	Moderate number of experts, generally control software engineers	Relatively very few trained experts in software security
Maturity of area	Mature	Mature	Evolving

The website links contained in this table were all accessed on July 15, 2012, except as indicated.

Risks of Security-Oriented Versus Safety-Oriented Software Systems

The distinction between security-oriented systems and safety-oriented systems has to do with which risks are emphasized in each. For the most part, as discussed previously, security is about protecting valuable and sensitive information assets from outside, and possibly inside, attacks, whereas safety relates to preventing the system from doing damage or causing injury to those using the system or otherwise affected by the system. The degree to which the software running within the system is responsible for either secure or safe operation of the overall system depends upon the specific functions that the software supports. Sometimes system security and safety are not dependent on software, but on physical isolation and the like.

It is also possible for a safety-oriented system to depend only on hardware to protect customers or other users from harm. Such physical interlocks are common for electromechanical devices. However, it is also feasible to design a system where software is given the responsibility for avoiding injury and the equipment is totally controlled by software.

Similarly, a security-oriented system can depend on physical methods for protecting assets, or, alternatively, the protection can be via software only. For example, electromagnetic media can have a physical switch or similar mechanism that prevents anyone from writing over existing content. On the other hand, the prevention of overwriting might be solely a matter of software controls.

For assuring some level of software security, information security professionals use methods such as threat modeling, penetration testing, and the like, to try to establish that outsiders or, to some extent, internal users cannot damage the system, misuse or steal data, and so on. While those responsible for software security pay attention to secure coding, functional testing and non-functional security testing, functional security testing (testing that the software system does not do that which it is not supposed to do) is mostly ignored as it is considered too difficult and costly. Unit testing, integration testing, and acceptance testing are done from the user perspective and focus on meeting functional requirements. The amount of nonfunctional testing performed is often affected by whether the basic system functionality works as specified, resources available for testing and time to market.

For software safety, emphasis is more on validation and verification, as described above, to ensure that a system does not malfunction or fail in a manner that will endanger human life or harm the environment. Those tasked with software safety pay much more attention to unit testing and integration testing than is done for software systems not having safety requirements. Generally, much more time and effort are allotted to ensure a high level of system safety than is expended on other forms of testing, such as security-related testing.

In many ways, the role of the security professional is a far more difficult one than that of the safety engineer. The former has to defend against all poten- (✳) tial attacks, whereas the safety engineer has to ensure that a finite population of malfunctions and failures of critial safety systems does not cause harm or loss of life to persons or other beings. That is to say, software security engineers often have an infinite number of possible attacks and vulnerabilities to deal with, (✳) whereas software safety engineers have a relatively limited number of possible scenarios, although that number could be very large.

Expertise Needed at Various Stages (APPENDIX C.)

Security expertise should be front-end weighted in the life cycle to ensure that security requirements are included and reviewed. The system design and coding practices need to be instilled from the beginning. They require expertise on secure application architecture and development practices. Security and functional expertise is needed in the testing and implementation phases. Details of (✳) levels of involvement for security-critical systems are included in Appendix C.

Safety expertise is also needed early to ensure safe design, and to ensure that safety requirements have been considered and included as appropriate. However, safe coding practices may not play as great a part if safety-oriented languages are used. Additional expert input is needed in the verification and validation stages. Details of levels of involvement for safety-critical systems are included in Appendix D. (✓)

Summary and Conclusions

It would appear that software safety engineers are better equipped to ensure that safety-critical systems meet stated industry and government requirements and standards in order to minimize the risk to human and other life, and the environment, of a failure of the systems.

Software security engineers have neither as stringent requirements nor the luxury of allocated time and resources to test for security, as do their safety software engineering counterparts.

As a consequence, while the development life cycles are similar in form for both security-critical and safety-critical systems, the emphasis at different phases is can be very different, as described in Appendix C and D, respectively.

Software security engineers generally appear to have a difficult time convincing their management that security risks are real and should be minimized. Another aspect is that protecting security-critical systems often requires considerable effort, resources and funding and, given the perceived risks and responsibilities for breaches, even minimal funding is often not forthcoming. Because of this huge gap, those interested in achieving high degrees of security for critical software are either presenting normative approaches that are unlikely to gain much traction, or hoping for magical clean-slate breakthroughs that will somehow change the game. The greater the problem becomes, the more it is apparent that the huge amount of effort—potentially costing in the trillions of dollars—is needed to produce secure systems going forward, replacing deficient systems where possible, and retrospectively fixing legacy systems that are unlikely to be replaced within years or decades.

It is possible that, as safety-critical and security-critical systems are integrated into systems of systems and cyber-physical systems, stringent safety requirements will impact security requirements because security breaches can readily lead to intentional or accidental malfunction or failure of safety-critical systems. The hope is that the impetus from safety requirements and standards will elevate security requirements and standards, rather than software security practices dragging down the safety-critical requirements and standards. It would seem to be a fair guess and a fervent hope that safety will become a major driver for improved software security.

Endnotes

[1]   Zelkowitz, M. V., "What Have We Learned About Software Engineering," *Communications of the ACM*, Vol. 55, No. 2, February 2012, pp. 38–39.

[2]   Parnas, D. L., Risks of Undisciplined Development," *Communications of the ACM*, Vol. 53, No. 10, October 2010, pp. 25–27.

[3]   Parnas, D. L., "Software Engineering—Missing in Action: A Personal Perspective," *Computer*, Vol. 44, No. 10, October 2011, pp. 54–58.

[4]   An early report on the initial impact of the *technical glitch* at Ulster Bank in Northern Ireland can be seen at www.bbc.co.uk/news/uk-northern-ireland-18530509, last accessed on July 15, 2012.

[5]   The February 13, 2012 report can be obtained from the FedBizOpps website at https://www.fbo.gov/index?s=opportunity&mode=form&id=8b857838c3d9237cafce59786bc0 5fb2&tab=core&_cview=1, last accessed on July 15, 2012.

[6]   See   http://www.nitrd.gov/fileupload/files/nitrdcybersecurityr&dthemes20100519.ppt, last accessed on July 15, 2012.

[7]   See DARPA-BAA-12-21 on "High-Assurance Cyber Military Systems (HACMS)" at https://www.fbo.gov/index?s=opportunity&mode=form&id=8b857838c3d9237cafce59 786bc05fb2&tab=core&_cview=1, last accessed on July 15, 2012.

[8] See DARPA BAA-12-17 on "Crowd Sourced Formal Verification (CSFV)" at https://www.fbo.gov/index?s=opportunity&mode=form&id=3b5cf23a978799579294399b332 68c99&tab=core&_cview=0, last accessed on July 15, 2012.

[9] See http://bsimm.com/community/, last accessed on July 15, 2012.

[10] Available via www.hsgac.senate.gov, last accessed on July 15, 2012.

[11] See http://commerce.senate.gov/public/index.cfm?p=Hearings&ContentRecord_id=a67 6548f-a2a7-40ff-a18d-889a7907801c, last accessed on July 15, 2012.

[12] See http://news.bbc.co.uk/2/hi/uk_news/383547.stm, last accessed on July 15, 2012.

[13] See www.newser.com/story/137703/oops-plane-mistakenly-warns-passengers-of-crash. html, last accessed on July 15, 2012.

[14] See http://www.thesun.co.uk/sol/homepage/news/4065489/British-Airways-Airline-accidently-play-crash-warning-twice-during-Miami-flight.html, last accessed on July 15, 2012.

[15] Gawande, A., *The Checklist Manifesto: How to Get Things Right*, New York: Metropolitan Books, 2009.

[16] U.S. Department of Defense (DoD), *Systems Engineering Fundamentals*, Fort Belvoir, VA: Defense Acquisition University Press, 2001. Available at http://www.dau.mil/pubs/pdf/SEFGuide%2001-01.pdf. Accessed July 15, 2012.

[17] *NASA Systems Engineering Handbook, NASA/SP-2007-6105 Rev1. Washington, DC: NASA (National Aeronautics and Space Administration)*, December 2007. See http://education. ksc.nasa.gov/esmdspacegrant/Documents/NASA%20SP-2007-6105%20Rev%201%20 Final%2031Dec2007.pdf, last accessed on July 15, 2012.

[18] Haskins, C. (ed.), INCOSE SE Handbook Working Group, *Systems Engineering Handbook: A Guide for System Life Cycle Processes and Activities, INCOSE-TP-2003-002-03.2.1*, San Diego, CA: INCOSE (International Council on Systems Engineering), January 2011 (revised August 2011).

[19] Office of the Chief Information Officer, DOE (U.S. Department of Energy), *Systems Engineering Methodology, Version 3: The DOE Systems Development Lifecycle (SDLC) for Information Technology Investments*, DOE G 200.1-1.A, September 2002, See http://energy.gov/sites/prod/files/cioprod/documents/SEM3_1231.pdf, last accessed July 15, 2012.

[20] U.S. Department of Defense, *Mandatory Procedures for Major Defense Acquisition Programs (MDAPS) and Major Automated Information System (MAIS) Acquisition Programs*, DoD 5000.2-R, April 2002. See http://www.acq.osd.mil/ie/bei/pm/ref-library/dodi/p50002r. pdf, last accessed July 15, 2012.

[21] Braude, E. J., and M. E. Bernstein, *Software Engineering: Modern Approaches, Second Edition*, Hoboken, NJ: John Wiley, 2011.

[22] Merkow, M. S., and L. Raghavan. Secure *and Resilient Software Development*, Boca Raton: FL: CRC Press, 2010.

[23] McGraw, G., *Software Security: Building Security In*, Upper Saddle River, NJ: Addison-Wesley, 2006.

[24] Christensen, M. J., and R. H. Thayer, *The Project Manager's Guide to Software Engineering's Best Practices,* Los Alamitos, CA: IEEE Computer Society, 2001.

[25] ISO/IEC/IEEE 12207-2008: Standard for Systems and Software Engineering - Software Life Cycle Processes. Available for purchase at http://ieeexplore.ieee.org/xpl/login.jsp?tp=&arnumber=4475826&url=http%3A%2F%2Fieeexplore.ieee.org%2Fiel5%2F44758 22%2F4475825%2F04475826.pdf%3Farnumber%3D4475826, last accessed July 15, 2012.

[26] Ferrell, T. K., and U. D. Ferrell, " RTCA DO-178B/EUROCAE ED-12B," in *Avionics: Elements, Software and Functions,* C. R. Spitzer (ed.), Boca Raton, FL: CRC Press, 2007, pp. 16-1–16-11.

[27] A good source for software V&V is NIST Special Publication 500-234, Reference Information for the Software Verification and Validation Process, as applied to healthcare software systems, available at http://hissa.nist.gov/HHRFdata/Artifacts/ITLdoc/234/val-proc.html, last accessed on July 15, 2012.

[28] Brenner, J., *America the Vulnerable: Inside the New Threat Matrix of Digital Espionage, Crime, and Warfare,* New York: Penguin Press, 2011.

6

Software Systems Security and Safety Risk

If you don't attack the risks, they will actively attack you. The real professional is one who knows the risks, their degree, their causes, and the action necessary to counter them, and shares this knowledge with his colleagues and clients."
—Thomas Gilb, *Principles of Software Engineering Management* [1]

Introduction

The word "risk" likely means something different to each and every one of us, judging from the many debates that arise whenever the topic is raised. Some well-respected researchers look to adopt simplified methods for assessing and managing risks, some think that the determination of risk is complex and requires advanced probability theory, and yet others believe that the whole effort should be disbanded because it is impossible to come up with accurate estimates of losses and of probabilities of those losses occurring.

What many researchers and writers on the topic of risk appear to miss is that the assessment of risk can be a very personal process. Risk mitigation may be done largely to reduce negative outcomes that affect individuals directly, even if someone is supposedly assessing risks that may be assignable to someone else or to other groups or organizations. There is a question as to the degree to which someone can be objective enough to assess another person's (or another entity's) risks. Clearly, actuaries are tasked with doing that job for their insurance company employers, but even then, they might be motivated, likely subconsciously, to bias estimates to favor their own careers. For example, if actuaries are conservative, payouts for losses will be less relative to premiums, but relatively high premiums will deter customers; if they are overly optimistic about catastrophic events not happening, the company could see large reductions in profits.

In this chapter, we examine the risk assessment and management processes, taking into account the many definitions and models that are offered. We also look at the motivations of risk assessors and how those motivations might affect the resulting analysis, and further how they may influence risk management more broadly. It will be shown that many assertions about risk miss the point entirely, which serves to explain the increasing number of inappropriate risk assessments and poor decisions which we have seen play out in recent years.

Understanding Risk

One might think from the seemingly knowledgeable manner in which self-anointed authorities discuss risk, that they thoroughly understand the concept and how to apply it. The results suggest otherwise, particularly when you consider the recent man-made catastrophes in the financial and nuclear-power worlds.

Risk determination and analysis have taken a number of major negative hits of late, particularly with respect to forecasting the global financial meltdown beginning in 2008, or predicting that a tsunami hitting northeastern Japan would not exceed a certain height, as it did on March 11, 2011, or anticipate that the Arab Spring would take place and spread in the manner that it has. The common thread of these events, and many other similar incidents, is that forecasts with respect to the likelihood that specific events, particularly catastrophic events, are often wildly off the mark. That is not meant so much as a criticism of our ability to predict the unpredictable as it is of recognizing that estimating infrequent catastrophic events is virtually impossible. For catastrophes, the best approach might well be to implement some broad-based monitoring, protective, and response measures that cover an extensive range of potential events, in the hope that the preparation will somewhat soften the blow and provide timelier and more effective responses.

Risks of Determining Risk

Risk is associated with taking chances; we take chances when outcomes of decisions or actions are not known in advance with certainty. This is the situation in which we find ourselves most of the time. Often the downside of decisions, which prove to be wrong or less than satisfactory, is not particularly damaging. For example, you may select a menu item in a restaurant and then notice that someone else has ordered a dish that you would have preferred. In this case, you can usually return to the restaurant at a later date and order the coveted dish for yourself. However, on other occasions, a bad decision can be disastrous and there may be no opportunity to recoup losses, unless the risk was transferred through insurance or other means. An example of this might be an explicit decision to build a house in a flood plain and take out flood insurance (assuming that insurance coverage is available for the contemplated location).

While there are opportunities to insure against attacks in cyberspace, such insurance can be difficult to come by because the applicant must demonstrate a strong security posture in advance for the underwriter to be comfortable enough with the level of risks to issue a policy and determine a reasonable premium to charge. One consideration when evaluating a company for cyber-security insurance is how the company builds and/or acquires safety-critical and security-critical software systems. It should be noted, however, that purveyors of software products generally assume no liability for vulnerabilities that might reside within the software or for the consequential damages that might be sustained. Mostly, acquirers of third-party software are contractually responsible for installing and operating the software correctly and assuming any losses that might occur through the use or misuse of the particular software product. Independent software vendors (ISVs) will typically only offer to refund the cost of the software product and will not assume any other liabilities.

Software-Related Risks

The quest to manage certain software-related risks was actively pursued in the 1980s. Dr. Barry W. Boehm, who is well-known for his definitive work on the software engineering process and famous for the spiral model [2], edited a pioneering collection of papers on software risk management [3]. At that time, researchers such as those who wrote chapters in Boehm [3] did not agonize over the precise definition of risk—their risk-related focus was on the software system design and development processes and their associated uncertainties. The challenge back then was to minimize uncertainties related to software development processes, rather than to protect against potential attacks on, or damage from, software systems after they have been developed and deployed. Nor did researchers examine, at that time, the risk of malicious developers, who might introduce destructive code into the software. Interestingly, even today the prospect of such threats is generally not considered, despite there being very significant risks associated with such activities.

During the computer program remediation effort to correct the Year 2000 problem of misinterpreting the year in date fields, large quantities of program source code were shipped to outsourcers, many of them offshore and many in

Asia, Eastern Europe, and Ireland. There was considerable apprehension among information security professionals that such third-party contractors would introduce "back doors" that is, computer-program code that would allow them to get into the software system at a later date and compromise the system in some manner or other for personal gain. There is little published evidence that this type of crime has in fact been perpetrated, although the likelihood of such an approach was thought to be high enough to voice concern publicly [4].

In the ensuing three decades, the role of software-intensive systems has become increasingly critical as such systems are being deployed into practically every area of human endeavor. Furthermore, the orientation of software-based systems has also changed, fueled by rapidly falling costs of electronics, leading to much cheaper systems and networks. There is also a burgeoning population of geographically dispersed developers, particularly in Asian, Eastern European, and South American countries, as well as in Russia.

The ubiquity of applications software products (apps), which are being created by hundreds of thousands of developers for Apple, Google, Facebook, and other platforms, has dispersed security, privacy, and safety issues throughout cyberspace. In previous chapters, we made a clear distinction between safety and security as the terms might be applied to software systems. To make matters even more complicated, the term *privacy* is also frequently subject to varying interpretations. Privacy involves the legal and social rights of individuals, with the implementation of security measures being one of the ways to implement privacy directives. Major security and privacy issues arise from applications software that is built by subcontractors for mobile devices, such as smartphones [5]. As smartphones and similar mobile devices are increasingly used to control physical systems, such as monitoring one's home, we are seeing safety issues coming to the fore [6].

These changes in the way in which commonly-used applications are developed, distributed, and used has meant that oversight and management controls regarding such development activities have become much more difficult to enforce, if even it is possible. Software-intensive systems produced in this open environment are much more likely to contain vulnerabilities and hence are easier to compromise. Consequently, while the pioneering work of the 1980s provides a good basis for mitigating risks stemming from the software system design and development processes themselves, they do not account for the large number of exploits that the Internet, mobile computing, and other technological innovations have produced. Nor did the early researchers anticipate the immense proliferation across the world of the development of software applications by individuals or small groups not subject to the discipline, oversight, and controls that can be expected from large software manufacturers. While some platform providers, such as Apple Inc., have in place some level of screening apps before they are made available to the public; such screening is mostly done to ensure that apps function correctly and do not contain malicious code or malware. This form of screening is less thorough when it comes to consideration of security, privacy, and safety consequences that might result from the large-scale deployment of these apps, which are typically used (and abused) by huge online populations that generate millions of uses per day.

Motivations for Risk Mitigation

A strong motivator for incorporating safety and security requirements into software systems is the identification, assessment, and mitigation of risks. In virtually all situations, a trade-off must be made between the quality of software (where quality comprises security, safety, integrity, resiliency and the like) and the time and effort that might be needed to get the software systems out the door within acceptable timeframes. The trade-off is more apparent for security-critical systems than it is for safety-critical systems; this is in large part due to the far greater consequences of a failure in safety-critical systems (such as control systems for avionics, transportation, electricity grids, oil rigs, space shuttles, and nuclear power plants), and a failure of information systems operated by banks, stock exchanges, payment card processors, and the like. It is perhaps due to this conflict in risks and goals that there seem to be so many deficient and defective software systems created, delivered, and operated, and many software development projects that are abandoned even before going into production. Of course, the press tends to report mostly data breaches and project and system failures. Understandably, the media do not report that software systems have been operating well and without incident and that they have met their expected performance requirements since this is not considered newsworthy, although given the complexities and challenges involved, it is a major achievement. Similarly, bringing in software development projects on time, on budget, and meeting all requirements where those requirements include the needs for security, safety, dependability, performance, availability, integrity, and resiliency are not commonly reported, although they do appear in industry publications from time to time.

Even though they are considered newsworthy, only a very small percentage of data breaches and software system failures are actually reported to the public. Major data breaches have indeed occurred, but have not been reported in the press or included in chronologies of data breaches. Affected organizations made no attempt to hide or cover up these breaches; these breaches were not reported on because only a relative few such events, compared to the total population of known and unknown breaches, are noticed by the press.

Some lack of reporting of privacy breaches, in particular, is also due to there not being notification requirements for such breaches in many countries, such as those in the European Union. In the United States, if data are encrypted, then unauthorized access to the data by hackers does not need to be reported; it is not required. This encryption criterion is somewhat misleading because the majority of attacks, estimated to be some 70 percent, are achieved through the application layer, in which case the attacker will be able to access the data since applications have to decrypt data in order to process them. Consequently, encryption has limited value with respect to protecting data if the applications processing the data are compromised.

Many U.S. state laws require notification to authorities, to the public at large, and to affected individuals in the event of data-security breaches that potentially affect nonpublic personal consumer and employee information. In Europe, although the original 1995 E.U. Data Protection Directive, which at the time of writing is likely to be superseded in 2012, was more stringent about protecting personal information than U.S. laws. The requirement for notification has been virtually nonexistent in Europe, for example, although that situation is in the process of changing.

Furthermore, many instances of data compromise are not detected by the specific organizations that are under attack by insiders or external forces. Often organizations are unaware that they have been successfully attacked until or unless some third party determines that an attack has in fact taken place because of consequences such as fraudulent activity, and the third party traces fraud back to the original breach. Because of the manner in which many breaches are in fact detected, it is fair to presume that the vast majority of breaches are never detected. In part, this is due to woefully inadequate data creation, monitoring, reporting, and response. In contrast, malfunctions and failures of systems controlling power-generation facilities, electricity grids, water treatment plants, airplanes, ground vehicles, and the like are usually very obvious because of highly visible consequences, such as train and airplane crashes or explosions at refineries, oil rigs, power stations, telecommunications hubs, chemical plants, and the like. Of course, even with control system failures, many near misses (for example, when the damage was avoided by human intervention, actual failures that did not not produce visible physical consequences, or failures that went unobserved) do not find their way into the press.

Defining Risk

The analysis of risk depends first and foremost upon one's definition of the term *risk*, which is not as obvious as it may seem. In this chapter, we shall examine a number of ways in which risk is defined and calculated. Once the approach to risk management has been determined, the implementation of the agreed-upon risk measurement methods then depends upon having certain information, often in the form of security-related and safety-related metrics, available to the analyst who is making the risk determinations. The results are then typically presented to decision-makers who are supposed to be in a position to make requisite trade-offs.

Furthermore, this chapter expands upon standard methods of calculating the return on security and safety investments in several ways. First, we take into account the dynamic nature of threats, vulnerabilities, and defenses. Next, we take a more holistic view of security and safety investments.

The protection of information assets can be viewed in two ways. One is the hierarchical view of security and safety measures, such as avoidance, deterrence, and prevention. The other is defense in depth, wherein various security tools and processes such as firewalls, identity and access management, and intrusion detection and prevention products are combined for greater overall protection and many levels of fail-safe procedures are introduced to prevent harm from coming to people or environments. The reader will gain a deeper understanding of the factors that affect risks and returns from investments in security and safety measures, tools, and processes, and will find that using a portfolio approach can lead to more cost-effective security and safety.

Assessing and Calculating Risk

There is an ongoing, frequently vituperative, and particularly confusing debate about what risk is, how it is calculated, who the appropriate decision-makers might be, how they should make decisions, and how those decisions should be implemented [7] Two thought-leaders in this area, namely, Donn Parker [8] and Scott Borg [9], have voiced very different opinions on risk assessment and the use of risk measures for use in determining one's information security program, in particular. While both Parker and Borg have concerns with current risk assessment and management practices, their issues differ. Parker has criticized the common approach to risk assessment, which uses estimates of damaging events and their likelihood; Parker claims that neither factor can be measured or estimated with any degree of accuracy, and so the entire approach should be jettisoned. He suggests that, in its place, the best of the practices used by peer organizations need to be established and adopted

Borg, on the other hand, supports the use of traditional risk assessment methods, but emphasizes that "[e]conomics drives security," and that the focus should be on "loss of value." He contends that the common fixation on assets is a mistake because the consequences of compromising a low-value asset, such as a cheap pressure gauge that gives false readings due to a cyber attack, can have huge consequences, such as the release of a toxic gas. This example emphasizes the need to determine the impact of consequences of failure rather than the intrinsic value of the protected asset.

In a couple of articles, Jeff Lowder questions a traditional approach [10] that is commonly based on the equation:

$$Risk = Threats \times Vulnerabilities \times Impact$$

where the term *consequences* (or *loss*) is often substituted for the term *impact*.

Lowder claims that such a formula is mathematically nonsensical and should be discarded. He believes that the formula should be expressed as:

$$Risk \text{ (Expected Loss)} = Function \text{ (Threats, Vulnerabilities, Consequences)}$$

That is to say, the expected loss from an adverse security or safety event is a function of exploits, rather than specific mathematical relationship with the attacks. Threats, in Lowder's presentation, are actual instances in which a threat has been realized as opposed to the possibility of an attack. The expected loss is further related to vulnerabilities that have not been mitigated and consequential losses in the event that an attack is successful. As Lowder correctly claims, the mathematical relationship is not multiplication, nor is it simple addition. However, giving those who support this formula the benefit of the doubt , one might assume that the use of addition and multiplication of the independent variables are not meant to be interpreted literally, but that they should only be used to show risk being dependent, in some fashion or other, on a series of specific independent variables. However, granting this assumption does not avoid the problem of the units of measure being inconsistent. Be that as it may, it makes much more sense and is also less confusing to be specific and rigorous in the use of terms, the expressing of relationships among variables, and the use of mathematical symbols.

As a further example of the imprecise multiplicative approach to the risk calculation, Borg [9] defines risk as follows:

$$Risk = Threat \times Consequence \times Vulnerability$$

where the independent variables are defined as follows (emphasis added):

- *Threat*: an estimate of the *likelihood* of a given type of security event in a given time period.
- *Consequence*: an estimate of the *losses* that could arise from the above security event.

- *Vulnerability:* an estimate of the *extent* to which the event will cause the above losses, given existing security measures.

The above equation and definitions of the independent variables somewhat reflect the common view of risk as a function of the probability of a loss (Borg's *threat*) and magnitude of the loss (Borg's *consequence*), which when multiplied together yields the expected loss. It would appear that *vulnerability*, as defined by Borg, might be considered to be a modifier of the probability of loss and can be expressed as a conditional probability (i.e., the probability of loss given some level of security in place). This allows for considering only those security measures that might reduce or eliminate the chances of a particular exploit actually doing damage. However, as suggested above, one could combine these definitions of threat and vulnerability into one factor, namely, the conditional probability that an attack is successful, given that a particular set of security measures have been put in place and are being maintained. In any event, given the appropriate units of measure, Borg's equation has some meaning reflective of the standard expected loss formulation.

A range of formulations for risk is shown in Table 6.1. One additional point to note is the difference between the negativistic view of outcomes leading to losses versus a neutral view that outcomes can be both losses and gains. From the insurance underwriter's perspective, only losses from negative events have a significant detrimental impact on costs incurred by insurers. If nothing bad happens, the insurers just continue collecting premiums with no payout.

A more detailed multistep process and somewhat different set of specific definitions and metrics is illustrated in Figure 6.1, which separates out the various individual influences of factors on risk.

Figure 6.1 shows the progression of threats to exploits, where only a subset of the full range of threats are actually realized in the form of exploits, and only a fraction of exploits have effective means of delivery and deployment. Of those threats that have been converted into exploits and have a means for launching, usually only a small percentage are able to penetrate defenses and get into systems. Once inside, there is always a question as to the real value of the assets at risk. It should be pointed out, however, that few assets obtained from random exploits have much value unless the attacker knows to target a specific set of data about which he or she is already aware and is able to transform into some type of benefit, usually financial fraud.

Threats Versus Exploits

There are many definitions of the term threat other than the one that Borg mentions in [9], in which threats represent the probability of adverse events. In their book on Microsoft's Security Development Lifecycle (SDL), Howard and

Table 6.1

Risk Assessment Methods

Approach	Calculation	Strengths	Weaknesses	Comments
Addition of variables	Risk = Asset + Threat + Vulnerability	Simple calculation Includes relevant variables	Units of measure not specified Values of independent variables are inconsistent Independent variables are not additive Meaningless calculation e.g., does not return zero if any variable is zero.	Presented, for example, in blog "Threat, vulnerability, risk—commonly mixed up terms" on Threat Analysis Group (TAG) website at www.threatanalysis.com/blog/?p=43 The blog states that "[r]isk is a function of threats exploiting vulnerabilities to obtain, damage or destroy assets." However, these variables are not additive.
Multiplication of variables	Risk = Threats × Vulnerabilities × Losses (or Impact)	Simple calculation. Risk calculated to be zero when one or more variables are zero.	Units of measure not consistent Numerical values of variables are inconsistent Independent variable are not multiplicative Meaningless calculation and results	See J. Lowder, "Why the 'Risk = Threats × Vulnerabilities × Impact' Formula is Mathematical Nonsense," available at http://www.bloginfosec.com/2010/08/23/why-the-risk-threats-x-vulnerabilities-x-impact-formula-is-mathematical-nonsense/
Multiplication of variables	ALE = ARO × SLE, where ALE = Annualized Loss Expectancy, ARO = Annual Rate of Occurrence, and SLE = Single Loss Expectancy	Includes variables for expected losses and probabilities of those losses	Period of one year is arbitrary—may have much shorter or longer horizons History is not necessarily representative of future events, especially when underlying assumptions change or when unanticipated events occur Focus on losses rather than expected values of outcomes generally	The ALE method is frequently touted as a good measure of risk, particularly for insurance purposes. For a definition and example of the calculation see www.riskythinking.com/glossary/annualized_loss_expectancy.php
Ordinal	Assignment of levels of importance, impact and likelihood, such as high, medium, low	The exercise itself can be very beneficial to the understanding of risk by various constituencies within an organization The OCTAVE® approach is generally accepted by many auditors, regulators, and the like	Yields a somewhat subjective view of the importance and potential impact of risk factors Subjective methods for combining risk factors Less meaningful when trying to aggregate risks across organizational units	A practical example of this approach is OCTAVE® (Operationally Critical Threat, Asset and Vulnerability Evaluation[SM]) developed by Carnegie Mellon University's Software Engineering Institute. See www.cert.org/octave/
Cardinal Scoring (i.e., numeric scores)	Assignment of scores, such as 1 through 10, to levels of importance, impact and likelihood	Allows for some ability to aggregate risks by department, company, etc.	Gives an unfounded sense of precision to the analysis and calculations	BITS Expectations Matrix, and FISAP (Financial Institution Shared Assessments Program), which became the Shared Assessments Program. See http://sharedassessments.org/

Table 6.1 (continued)

Approach	Calculation	Strengths	Weaknesses	Comments
Expected value	Calculation of risk as the product of a particular outcome (gain, loss, neutral) and the probability of its occurrence	Incorporates the likelihood of an event into the calculation, requiring responder to estimate the chance of occurrence Allow for gains as well as losses	Probability of occurrence is highly subjective Does not use probability distributions—just uses a single point estimate	See, for example, J.A. Jones, *An Introduction to Factor Analysis of Information Risk (FAIR)*, Risk Management Insight, available at www.riskmanagementinsight.com/media/docs/FAIR_introduction.pdf
Expected loss	Same as above except that it is biased towards losses, which derives from the insurance industry perspective	As for expected value, above, except that gains are not contemplated	As for expected value above, except that positive outcomes are not included because insurers only need to respond to negative outcomes	For a treatment of expected loss, see D.W. Hubbard, *The Failure of Risk Management: Why It's Broken and How to Fix It*, Wiley, 2009.
Probability distributions	Uses probability distributions rather than point estimates of likelihood of occurrence and impact	Provides a more realistic view of how the estimates of the probabilities of occurrence and impact of risk factors vary	Much more difficult to get accurate probability distributions from estimators Some do not believe that this is meaningful or doable	For methods to obtain probability distributions, see D.W. Hubbard, *How to Measure Anything: Finding the Value of Intangibles ion Business, 2nd Edition.* Wiley, 2010
Objective	Attempts to remove subjectivity by asking subject-matter experts (SMEs) to provide estimates	Removes some of the bias introduced by belonging to a group that will be directly affected by results of the analysis and/or by the impact of adverse events	May not benefit from input by those who are "in the line of fire" Subject-matter experts will likely have their own biases based on their particular experiences	See C.W. Axelrod, "Risk Mismanagement – Scoring vs. Monte Carlo vs. Scoring," on September 12, 2011 at http://www.bloginfosec.com/2011/09/12/risk-mismanagement-%e2%80%93-scoring-vs-monte-carlo-vs-scoring/
Personal	Recognizes that there is always some measure of bias and accounts for personal viewpoints	Provides more realistic view of how the estimates are derived	Extremely difficult to remove personal bias in order to come up with an objective view	See C.W. Axelrod, "The Personalization of Risk," on December 19, 2011 at http://www.bloginfosec.com/2011/12/19/the-personalization-of-risk/
Holistic	Includes all aspects of risk from personal, subjective and objective sources	Accounts for all risks, e.g., internal, external, known, unknown, etc., and all the viewpoints of all assessors, e.g., personal, subjective and objective	Difficult to identify and get acceptance of some risk factors Difficult to obtain data for some risk factors Extremely difficult to weight and aggregate risks over all entities and constituencies	

All the links in the above table were accessed on July 18, 2012.

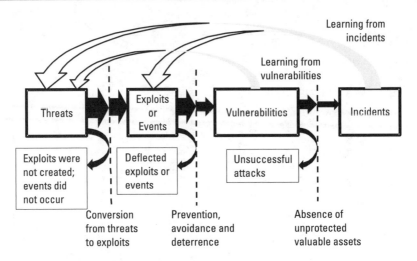

Figure 6.1   Threats, exploits, events, vulnerabilities, and incidents.

Lipner [11] define a threat as an *attacker's objective*, while noting that some view a threat as the actual attacker or adversary.

These definitions agree broadly with the model illustrated in Figure 6.1, in which a threat is thought to embody the *potential* for, or *likelihood* of, developing a working *exploit*, but is not considered to be the exploit itself. It should be noted that, in this model, a threat is really a prerequisite for an exploit and is not something that might lead directly to a security or safety incident. The exploit has still to overcome several hurdles before it can be credited with having triggered an attack.

In Figure 6.1, an exploit is shown as the functioning realization of the threat's intent, as it were—typically a piece of malicious computer code (commonly termed *malware*) that acts upon a target system for the purposes of engaging in some form of illegal, fraudulent, or otherwise damaging activity. That is to say, a threat, as so defined, is not in and of itself a danger—it is the precursor of an activity that could lead to a dangerous product in the form of an exploit.

A threat can also be construed as the initial step towards a potential incident, which could be intentional or not, but which will likely damage or compromise a system nonetheless. For example, there is always a threat that software contains errors or bugs that could result in malfunction or failure, or that operational errors might cause damage. There are unplanned and unintended exploits leading to incidents, such as when a circuit component burns out, which might cause the system to fail. This means that there are only a (likely) small percentage of threats that are ever converted into exploits, and fewer still when one considers the small fraction of all exploits that are released by their creators and are also successful in their attacks of their targets. Also, the impact on the software system of an error by a developer, operator, or some other individual may be reduced by designing resiliency into the system using technologies that are self-healing, fail-safe, and/or redundant, as described in Axelrod [12].

Whereas one might say that this approach validates Borg's threat definition, in that we should end up with the aggregated likelihood that a particular security event or incident occurs, we get far more insight into the process, and likely more accurate estimates of the probability of successful attacks, by using the more multistep model illustrated in Figure 6.1, particularly if we extend the approach to include consideration of time-dependency (or dynamics) and the human behavior of stakeholders, as we do later in this chapter. Also, the combination or aggregation of different types of risk (or of the same risk assessed by different individuals or groups) is not straightforward, particularly when one includes probability distributions that are usually more appropriate, though more difficult to collect, than are single point values.

Many approaches to risk assessment require only point estimates of the loss due to an incident and of the probability of that loss. However, such factors are more accurately expressed as probability distributions. For example, the loss might follow a normal or lognormal probability distribution with mean μ and standard deviation σ. While it is difficult enough to get assessors to come up with point values, it is considerably more demanding for those doing the estimating to come up with probability distributions for particular variables. Furthermore, while estimates of point values for variables will vary among individuals, requests for probability distributions of these values will add significantly to the data-collection process, and the results from such requests are much more difficult to reconcile, analyze, and aggregate.

Typically, we have to consider trade-offs among more realistic models, the cost of collecting and analyzing the more complex data, and the impact that more detailed models will have on decisions. If decisions are not improved by adding this layer of complexity, then requiring the probability distributions may not be worthwhile. However, if there is expected to be a significant improvement in decision-making, and if it is likely that the more complex approach will lead to better decisions, where better means reducing risk significantly, then it might make sense to put in the extra effort.

Threat Risk Modeling

There is some disagreement, and possibly considerable misunderstanding, when it comes to the definition of threat modeling or threat risk modeling. In terms of the model shown in Figure 6.1, what is being modeled in most cases are exploits rather than threats because writers have provided specific instances of threats that have been realized as an attack or exploit, yet still refer to them as threats. This confusion of terms carries through to threat modeling. Perhaps a more meaningful term would be *exploit risk modeling.*

Threat risk modeling is basically the development of models that can be used to try to anticipate the range of exploits that might be directed against a software system's vulnerabilities, the likely impact that would have been felt if no preventative actions had been taken, and the anticipated impact if various protective and avoidance actions are taken. The processes described below provide a good background in how to go about performing threat risk analyses, but they do not guarantee that all potential exploits and existing vulnerabilities will be accounted for, especially in the intensely dynamic world of attacks against, and defenses of, mission-critical software systems. Performing such risk analyses requires both knowledge of the changing and intensifying threat environment and a detailed understanding of the components of the software systems (applications, system software, firmware, hardware, interoperations, etc.). Because few individuals have such a broad knowledge, except perhaps for the smallest of software systems, it is necessary to bring together subject-matter experts, usually including specialists from third-party service providers and vendors, to address the threats, vulnerabilities, and mitigation strategies for the typical large, complex software systems found in today's organizations.

McGraw [13] claims that Microsoft employees use the term threat modeling when they really mean risk analysis, McGraw quotes Swiderski [14]: "During threat modeling, the application is dissected into its functional components. The development team analyzes the components at every entry point and traces data flow through all functionality to identify security weaknesses."

Despite McGraw's contention, Microsoft employees Howard and Lipner include their discussion of threat modeling [11]. They negate the claim that the term *threat modeling* is misused at Microsoft. Just to confuse matters further, the Open Web Application Security Project (OWASP) straddles the issue by using the term *threat risk modeling* in its description of this activity [15]. Incidentally, OWASP endorses Microsoft's approach to the threat risk modeling process, which, according to OWASP, consists of the following five steps:

1. Identification of security objectives;
2. Review of the application;
3. Decomposition of the application;
4. Identification of threats;
5. Identification of vulnerabilities.

Howard and Lipner [11] list nine steps in the process, namely:

1. Definition of use scenarios;

2. Constructing a list of external dependencies;

3. Definition of security assumptions;

4. Creation of external security notes;

5. Creation of one or more data flow diagrams (DFDs) of the applications being modeled;

6. Determination of threat types;

7. Identification of threats to the system;

8. Determination of risk;

9. Planning of mitigations.

Clearly there are parallels between the two processes, although Microsoft's list is more comprehensive than the OWASP list.

Perhaps the most interesting aspect of the threat modeling process, as it relates to our discussion, is the classification of threat types by Howard and Lipner [11], using the acronym STRIDE to identify types of threat. STRIDE stands for the following categories of security risk:

- *Spoofing identity:* One user must not be able to assume the attributes of another user.

- *Tampering:* Users must not be able to modify data or program code in unauthorized ways.

- *Repudiation:* Users must not be able to deny having done something that they actually did.

- *Information disclosure:* Users must be prevented from disclosing sensitive information.

- *Denial of service:* The system must protect against attacks that cause the system not to be available to authorized users.

- *Elevation of privilege:* Users must not be allowed to gain increased capability beyond what they are authorized to perform.

In the above definitions, users might be considered to be valid (i.e., they might have been authenticated and duly authorized) with specific rights with respect to particular software systems, or they could be malevolent outsiders or insiders with the intention of doing harm. However, it should be recognized that both valid and malicious users often exhibit similar usage patterns because the hackers frequently masquerade as authentic users, as in identity spoofing. Furthermore, access rights and privileges are very often beyond what legitimate users need to know because identity and access management for many large-scale systems might be poorly managed. Assigning application access rights to users is complex and is further complicated by user populations that are constantly changing through internal reassignments and external events, such as natural disasters, takeovers, or mergers.

Threats from Safety-Critical Systems

When it comes to safety-critical software systems, which are frequently industrial or military control systems, threats are *from*, rather than *to*, software systems, as we have discussed. Now that control systems are increasingly being attached to networks, particularly public networks like the Internet, control systems are becoming subject to many of the same security threats that plague security-critical information systems.

Safety-critical software-intensive systems also have to tackle their own set of threats and exploits. Even if the networks supporting control systems are physically and logically isolated, it is still possible to enter the network via other vectors, such as USB drives, as was used for the Stuxnet worm. Had the victim of Stuxnet, a uranium-processing system in Iran, been completely closed, the particular method used to infiltrate the system would not have been viable. Futhermore, if the Stuxnet virus had not "escaped" it would not have replicated over the Internet; it may have remained a secret. It is claimed that the virus got out into the world because of a programming error [16].

The consequences of malfunctions or failure are used to determine the level of safety that must be achieved for particular certification. In Tables 6.2 and 6.3, we show the various levels attributable to control systems and information systems in aircraft (i.e., avionics) and in automotive vehicles, respectively [17].

It should be noted that, for the most part, control system malfunctions and failures are considered to be much more damaging than failures of information systems because the former could result in physical injury or loss of life, whereas the latter usually cause inconvenience, annoyance, and possibly loss of money and other assets (i.e., intellectual property). Nevertheless, secondary and tertiary effects of an information-system failure could impact the physical world. For example, the compromise of a financial system might result in major financial losses, which could lead to damage to one's lifestyle or worse. For example, the month-long computer system outage of the main processing systems of Ulster Bank in Northern Ireland, which lasted from mid-June to mid-July 2012, caused a great deal of hardship to customers, even though all of the customers will be compensated for any resulting monetary losses.

Of course, malfunctions and failures can be internal to safety-critical control systems, as would result from programming errors, or they can be induced by unauthorized access to those systems and consequent attacks upon them.

Table 6.2
RTCA/DO-178B Standard Applied to Aircraft Certification

System	Type of System	Level A (Catastrophic)	Level B (Hazardous)	Level C (Major)	Level D (Minor)	Level E (None)
Flight control system	Control	X	—	—	—	—
Cockpit display and controls	Control	X	—	—	—	—
Flight management system	Control	X	—	—	—	—
Brakes and ground guidance system	Control	—	X	—	—	—
Centralized alarms management	Information	—	—	X	—	—
Cabin management system	Information	—	—	—	X	—
Onboard communications system	Information	—	—	—	X	—
Centralized maintenance system	Information	—	—	—	X	—
Entertainment system	Information	—	—	—	—	X

Table 6.3
IEC 61508 Applied to Automotive Vehicles

System	Type of System	SIL 3	SIL 2	SIL 1	Not Applicable
Steer-by-wire	Control	X	—	—	—
Brake-by-wire	Control	X	—	—	—
Engine management system	Control	—	X	—	—
Dashboard	Information	—	—	X	—
Body controller	Information	—	—	X	—
Navigation	Information	—	—	—	X
Diagnostic	Information	—	—	—	X
Entertainment system	Information	—	—	—	X

Note: SIL = Safety Integrity Level [18].

The damage caused by the malfunctioning or failure of security-critical information systems is usually less dramatic than for safety-critical systems because compromises usually involve financial losses rather than physical harm as mentioned above. Nevertheless, there are many parallels with safety-critical systems. It is just that, to date, no one has categorized the severity of the damage caused by malfunction and failure for information systems in a comparable way to those categories used for certifying safety-critical systems because there are no established standards against which such information systems might be certified. The losses for both security-critical information systems and safety-critical control systems can be expressed in monetary terms, although the estimates for the latter will tend to be less accurate because it might involve putting a value on loss of life or injury or damage to the environment; both valuations are highly subjective.

In the financial services industry, there have been some attempts by regulators and industry bodies, such as the Payment Cards Industry (PCI) Security Standards Council, to establish criteria and standards to be applied to critical systems, such as those handling sensitive personal information or those running core processes vital to the continued operation of financial systems. The failure of Ulster Bank's automated processing systems on June 16, 2012, underlines this need.

There is a pressing and urgent need for such categorization and certifications in the security-critical information systems arena along the same lines as is currently done for safety-critical systems.

In Figure 6.2, we illustrate the types of attacks (such as external/internal, intentional/accidental) that might occur on security-critical and safety-critical software systems and what the consequences of successful attacks and malfunctions might be.

Creating Exploits and Suffering Events

There are many existing and potential threats that cannot be, or will not be, converted into exploits that can be used to attack software systems. Conversely, there are many events that occur spontaneously or are the result of human error or ineptitude.

Many exploits are unintentional (due to software errors, operational mistakes, etc.) or due to natural disasters, although the latter occur mostly in the physical domain. There is always the threat of human error, and it needs to be accounted for in ensuring that software systems are safe and secure. On the other hand, there are many occasions when human action averts the impact of attacks. This can occur before these threats are realized, at which point they actually become dangerous exploits, or when they are meting out their damaging payloads. For example, manual overrides are frequently built into industrial

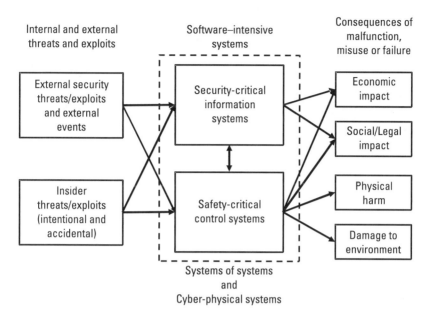

Figure 6.2 Threats and exploits and the consequences of system compromise and failure.

control systems. If the system goes haywire due to an error or attack, or if there is a natural disaster (such as a hurricane, earthquake, tsunami, etc.), operators can shut down systems, switch over to backups, and perform other mitigating actions.

With respect to the creation of real-world exploits, one reason for not developing fully-fledged exploits based on known vulnerabilities might be that the technology to implement the exploits does not exist, is not feasible, or is in the process of being developed but is not yet available. Another reason that threats are not turned into exploits might be that an exploit may be too complex or too expensive to develop and launch. In other cases, a hacker may not think that potential gains from a successful exploit are worth the effort, expense, or personal risk. In yet other cases, an exploit might indeed be feasible, but the developer of the exploit may decide not to release it, but to sell it to buyers who want to keep it "under wraps" and use it when and where they wish. Examples of such buyers might be makers of the software against which the exploit could be targeted, competitors of vulnerable software who are interested in ensuring that their own software is protected, nation states and terrorists wanting to keep a particular exploit against a specific vulnerability in hand for use in cyber warfare, organized criminals who are looking to attack specific public and private-sector organizations for financial gain, and the like. Here, vulnerability is thought of as a weakness that can be exploited by attackers in their quest to obtain financial or other gain, or perform acts of damage and destruction.

Unanticipated and unauthorized events happen frequently. They may result from errors in programs or from the manner in which the software systems are installed and used. Errors may be introduced at any point in the development life cycle, including during distribution, installation, and operation. Howard and Lipner [11] present the estimates of the costs of fixing such errors based on when the defect is introduced and when it is found, as shown in Table 6.4, which is derived from a book by Steve McConnell [19]. McConnell derived the table from eight other sources.

Some of these relative costs appear to be inconsistent. For example, one would expect that the cost of fixing defects introduced in the requirements phase and found during the test phase would be greater than those introduced in the architecture phase and found in the test phase, and even greater still than those introduced in the construction phase and found in the test phase. Similarly, it would be expected that the cost of fixing defects found once the software is in production would be greater with respect to the range of values for those defects introduced in the requirements phase than those introduced in the architecture phase. Despite these apparent anomalies, the statistics clearly show that, in general, the earlier a defect is introduced and the later it is discovered, the more expensive it is to fix.

The remediation cost matrix has significant impact on whether or not an organization decides to take the hit from known exploits or make the effort to fix the defect or vulnerability that would allow the attack to be successful. Clearly, as the cost of remediation is so much larger for defects introduced early in the life cycle but not identified until much later, it is less likely that the fix will be done late in the life cycle unless the losses are expected to be very large. In other cases, companies appear to be willing to sustain losses that are considered manageable and a cost of doing business [20]. Sometimes, if feasible, it pays to replace a software-intensive system altogether rather than trying to fix it, although only too often organizations choose to throw good money after bad and keep the old systems running for as long as possible. At some point in the lifespan of IBM's OS/360 operating system, it was found that trying to fix

Table 6.4
Relative Costs of Fixing Defects at Various Stages of the SSDLC

Phase where Defect was Introduced	Phase Where Defect was Found				
	Requirements	Architecture	Construction	Test	Production
Requirements	1	3	5–10	10*	10–100*
Architecture	N/A	1	10	15	25–100
Construction	N/A	N/A	1	10	10–25

*Inconsistent values based on expectations with respect to the form of the relationships.

defects actually resulted in introducing more bugs than were fixed, suggesting that it was no longer worthwhile to put in fixes and that a complete rewrite was justified [21].

When an exploit, say in the form of a computer virus or worm, has been developed but not released it is considered to be "in the zoo," whereas malware that has been activated and released is characterized as being "in the wild." As long as an exploit remains in the zoo, it does not endanger potential victims, but its very existence does present a risk because the exploit could well be released accidentally. A recent example of this concern was the attempt to prevent publication of a paper describing a mutation of the avian-flu virus that was developed in the laboratory. There was great concern that the virus could be accidentally released or that terrorists or other evildoers might use the technology to develop the virus for their own use [22].

Of course, an in-the-zoo exploit may or may not actually work. Often, hackers will run tests on early versions of malware to determine their potential effectiveness without actually releasing them. An astute security professional who also has access to a wealth of network traffic data at the transaction level might be able to spot the prototyping of exploits, as has Dr. Ed Amoroso, the chief information security officer at AT&T. In other cases, such as the notorious Morris worm, the exploit got out of hand and did far more damage than its creator, Robert Tappan Morris, had intended [23].

Vulnerabilities

Figure 6.1 shows that some percentage of exploits is released into the wild or directed against specific targets or set of targets. However, successful attacks depend on vulnerabilities existing and being accessible or exploitable by the attacker. McGraw [13] defines a *vulnerability* as "an error that an attacker can exploit." McGraw goes on to state that vulnerabilities arise from defects, which are called *bugs* at the implementation level and *flaws* at the design level. However, it should be noted that software might contain few, if any, discernible defects, yet still be subject to compromise. For example, Professor Ed Felton and his team discovered that by cooling down memory chips, data was retained beyond when the power to the chip was cut off, enabling researchers to pull information off the chip before it disappeared [24]. This example underlines the importance of context. It is not sufficient to build application software products that are demonstrably free from defects; the software system needs to be immune from attacks against operating systems, and against the logical and physical platforms upon which it might reside.

Of course, known vulnerabilities can be fixed in many cases, usually by installing patches issued by the maker of the software. There are, however, many vulnerabilities for which patches are not available or are not known, possibly because information about them is restricted or has been sold to the highest bidder. Nevertheless, given the existence of an exploit that can be directed against a vulnerability, there is a good chance that a hacker will attempt to use the exploit against an unprotected software system and be successful in gaining unauthorized access, inserting malware into the system, copying sensitive data, modifying programs, and the like. However, there is one last defense, namely, that the intruder—be it a person, an application, or a transaction—does not find anything of value to steal or otherwise compromise. This is an example of avoidance.

Responses to attacks can be active or passive. They can be achieved not only by protecting assets against attacks, but also by avoidance and deterrence, as shown by one of the curved arrows in Figure 6.1. Deterrence requires that victims of attacks or their representatives, such as law enforcement, announce that they will take effective legal action against perpetrators who are found and apprehended, and are seen to do so. However, because of the ability of hackers to attack from practically anywhere in the world with relatively little fear of being apprehended and punished, deterrence is often not very effective, particularly in a global context. Avoidance has a lot more going for it; if critical systems are not readily accessed by outsiders and restricted to insiders on a need-to-know basis, then chances are that unauthorized access will be minimized. Additionally, if care is taken to avoid putting sensitive data within easy reach, the effectiveness of attacks will be further diminished or eliminated.

Application Risk Management Considerations

The OCC Bulletin 2008-16 states that national banks should ensure that the manner in which applications are developed and tested addresses risks to the confidentiality, availability, and integrity of data, particularly customer information [25].

While the OCC Bulletin 2008-16 states that scope may vary with the size and complexity of a bank, it asserts that the following key factors impacting risk should be considered:

- Accessibility of the application via the Internet;
- Processing or provision of access to sensitive data;
- Source of development;
- Extent of secure practices in application development;
- Effective, recurring process for monitoring, identifying, and fixing vulnerabilities;

- Periodic independent assurance of security of applications;

- Management responsibility for ensuring applications are secure at acquisition (i.e., in RFIs and RFPs) and thereafter;

- Evidence of vendor adherence to sound practices and validation through third-party testing and/or audits for higher-risk applications;

- Support of purchased applications via vulnerability identification and remediation processes;

- Assurance that purchased and contracted applications are subject to the banks' ongoing testing process.

For in-house developed applications, OCC Bulletin 2008-16 suggests that banks need to include the following in their enterprise-wide efforts:

- Incorporation of appropriate attack models;

- Analysis of applications' environments;

- Ensuring that open source applications are subject to appropriate development and assurance processes;

- Assurance that personnel are trained in and aware of risk in the IT environment;

- Ensuring the appropriate protection of transactions and customer data by engaging in periodic application testing or validation.

While directed primarily at U.S. banks, these considerations should be applied generally.

Subjective vs. Objective vs. Personal Risk

A column "Risk Mismanagement—Scoring vs. Monte Carlo vs. Scoring" triggered a discussion as to the difference between subjective and personal risk postures [26]. Comments by Douglas Hubbard in response to the column prompted thought about the difference between subjective risk assessments and personal risk assessments. It is argued that the term subjective means a person's individualistic view of a risk relating to someone or something else, which may or may not cause the person assessing the risk to change actions and activities. On the other hand, personal risk assessment is about the direct impact of the decision on the person making the risk assessment and subsequent decisions.

Personalization of Risk

The value assigned by someone to a particular risk is usually very personal. As Jones [27] points out: "… risk tolerance is unique to every individual." That is, individuals generally assess risk in large part with respect to how they might be personally affected by a particular adverse or beneficial event and its outcome. Jones and others recognize that risks are personal or individual, but they do not suggest how to address this most significant, least understood, and least accounted-for factor in the whole of risk management. Risk-assessment values should be adjusted to allow for the personal and subjective biases that are commonly introduced. This is an area of research unto itself and cannot be dismissed with a simple acknowledgment that it exists. It may well be that a method, similar to the interview approach propounded by Hubbard in [28] for obtaining estimates of impact and their likelihood of occurrence, might be applied to the issue of the personalization and subjectivity of risk factor estimation.

That is not to say that decision-makers do not account for the adverse or positive impact of incidents on others. As mentioned previously, an actuary at an insurance company will likely determine the probability of loss events and the potential losses incurred in as objective a way as possible to provide his or her employer with the basis for underwriting specific risks of loss and determining the amount charged in premiums. On the other hand, that same person will decide whether or not to buy an insurance policy based on his or her personal view as to the likelihood of incurring such a loss and the degree to which the actuary might be personally affected; the actuary's risk assessment is not likely to be absolutely objective. The actuary may well exhibit some bias related to the need to be conservative so as to not subject the insurance company to excessive risks, as well as the desire to advance one's career by being recognized for thorough analysis and accurate forecasts. Further, there will also be some level of subjectivity based on the actuary's prior successes and failures.

The Fallacies of Data Ownership, Risk Appetite, and Risk Tolerance

In much of business and government, it appears that if you know the buzzwords and what the acronyms stand for, you are considered a *bona fide* expert in a particular field. This seems to be the case in the risk area, where, if you know what the acronym GRC stands for (yes, it's governance, risk [management] and compliance), then you are clearly one of the cognoscenti. You are then able to talk convincingly about "data owners," "risk appetite," and a host of other equally meaningless terms. At the risk of having riled up many of my readers, I will try to support these wild claims.

Table 6.5
Comparison of Influences on Risk Assessors Based on Level of Objectivity

	Objective	Subjective	Personal
Influences	Ability to demonstrate that correct risk approach was taken	Depends on expertise and experience and concerns about acquiring and retaining customers	Depends on self image and concerns about personal damage
Examples	Actuary using historical statistics and predictions of weather patterns, global warming, etc., to derive estimates of flood-related risks over a broad geographical area	Insurance broker expressing to potential customers the need for insurance to cover the risk of flood in a particular area	Insurance broker concerned that his or her own home might be flooded

I have sat through too many meetings where "the team" has worked its way through a long list of data items and assigned "owners" to each one. An example is "Social Security number" or SSN, which might logically "belong to" the head of the New Accounts Department, but it is unlikely that the person in question really tracks the use of the SSN. Typically, the SSN weaves through many systems within an organization, and so it is questionable as to whether the owner really understands what underlies the formal requests for approval for using SSNs in other applications. In addition, it is unlikely that the data owner is fully aware of the legal and regulatory requirements for protecting such information and the ramifications if the data should fall into the wrong hands.

This leads us to ask, what does "ownership" entail? Well, it means that the owner (whoever "owns" the data item, like a Social Security number or a password) is responsible for ensuring the security and integrity of that data item, that it is only made available on a need-to-know basis, that it is suitably protected (e.g., encrypted) whether in transit or at rest, and that when it no longer serves any useful purpose, it and all of its instances are appropriately deleted and destroyed. That is all well and good, but are the processes in place to supply the data owner with all of the information required to make decisions? Do computer software developers meet with the data owner and carefully describe how the data will be used, who will see it, who can change it, and who ensures that there are no stray instances of the data when the data files are destroyed? Frequently, many of these decisions are made by the information security staff in collaboration with an administrator. They act as surrogates for the true owner of the data who may be the only person that truly understands the importance and sensitivity of the data, if in fact he or she does so. Is it any wonder that so much highly sensitive information is inadvertently leaked?

We then have *risk appetite* (also called *risk tolerance*). It is incumbent upon managers to express their feelings about the risks surrounding a particular decision. But whose risks are being presented? The decision-maker will likely mix together a whole range of risks from those that are highly personal to those about which they do not really care, since those risks might not be seen to affect the decision-maker. Part of this characteristic is determined by whether the risks involve the decision-maker's own funds versus those of the institution, whether the risks are seen as personal versus corporate, and the chance of something negative or positive occurring "on their watch." Jones [27] points out that "[r] isk analysis only identifies how much risk exists" and not what level of risk is acceptable or unacceptable.

Then there is a whole range of risks for which nobody appears to have responsibility, although the consequences of negative events might well affect a broad range of stakeholders. This is referred to as the *tragedy of the commons*, as described in Hardin [29]. Generally, when there is no owner of a risk, such as the security of the Internet, the outcomes of failure of those resources for which no one takes responsibility are much worse than would have occurred if there was specific responsibility for those resources.

The Dynamics of Risk

Risks related to software-intensive systems are not static and they do not change according to any prescribed schedule. Different risk characteristics change in response to both external and internal factors. It is important, therefore, to account for such changes, often as quickly as they occur or are discovered, or preferably in anticipation of new exploits, the introduction of new features and new releases of software products, or changes in business volumes and mix.

With respect to the introduction of new features, there is always a risk that a new feature will itself have unintended consequences or that, in combination with other applications and software-intensive and data-intensive systems, a feature or group of features will provide unexpected capabilities that may contain undesirable aspects. A current example of this is the proliferation of apps that facilitate many activities but that also collect substantial amounts of personal data. This data, especially when combined with data from other sources, can reveal much more information about individuals than they would have agreed to if they had understood the potential for compromising their privacy. A good analysis of the risks of *big data* is provided in Angwin [30]. Axelrod [31] predicts that there will be a major shift in emphasis from issues relating to the security of data and software systems to privacy issues, and it appears from the increased press and rising concerns expressed in professional publications that this is happening. However, it is also common to see new, as well as longstanding features, being exploited by hackers to gain access to systems and data.

Regarding new releases of software, it was well-recognized as far back as the 1970s that, as software became more complex over time, the upgrade process reached a point of diminishing returns. In their definitive article, Belady and Lehman [32], who were directly involved in the maintenance and enhancement of the IBM OS/360 operating system, present three "laws" relating to the dynamics of systems and their need for continual effort and attention.

The demand for continuous repair and improvement is expressed in Belady[11] as the "First Law of Evolution Dynamics" as follows:

"I. *Law of continuing change.* A system that is used undergoes continuing change until it is judged more effective to freeze and recreate it."

A system will degenerate over time and become less manageable and more complex as it is changed, as expressed in the "Second Law of Program Evolution Dynamics," as follows:

"II. *Law of increasing entropy.* The entropy of a system (its unstructuredness) increases with time, unless specific work is executed to maintain or reduce it."

The "Third Law of Program Evolution Dynamics" reflects how a system and the process for developing and maintaining it, is "constrained by conservation laws," and is expressed as follows:

"III. *Law of statistically smooth growth.* Growth trend measures of global system attributes may appear to be stochastic locally in time and space, but, statistically, they are cyclically self-regulating, with well-defined long-range trends."

These laws describe how the structure of a system inevitably degenerates over time and requires substantial effort to maintain a desirable level of operability. At some point, it might not be worth the effort to try to keep on improving existing software systems, at which time the software maker has to decide whether to continue supporting the existing system or to replace it with a new system.

However, creating a replacement system also involves risk. The reliability of new software systems is lower at the point of introduction of the system and improves over time until a new version or an upgrade is released, when the reliability drops again. In contrast, when large complex software systems undergo major revisions, the malfunctions and failures may in fact increase over time and may not drop back to meet the prior failure-rate trend as would be expected if the system were replaced.

A Holistic View of Risk

Information security and control-system safety professionals tend to take somewhat narrow views of their fields and their responsibilities. The challenge of determining risk is to engage all the various stakeholders and somehow account for their personal and subjective views about the risks relating to a particular software system. The vast majority of publications about software security and safety are written by those who have a strong tendency to emphasize broad-based technical approaches rather than the specific functionality of particular software-intensive systems.

The willingness to invest in security and safety is highly dependent upon whose point of view is being taken. These viewpoints will vary significantly among stakeholders, with the emphasis on one or the other often depending on the relative authority of each group of stakeholders as well as the degree to which losses might be incurred by each of the various groups. While it is much more difficult to take this approach, the value produced and the reduction in friction will often overwhelm the costs.

Summary and Conclusions

We have considered a broad range of approaches to risk assessment and risk management, pointing out the deficiencies of some and expressing support for others. In general, the more simplistic approaches, which are based on point estimates, ordinal (e.g., high, medium, low), or cardinal (e.g., 1 through10) measures, tend to be generally representative rather than precise measures; yet, followers of these approaches are often encouraged to take the results as more accurate than the approach warrants. Risks should be determined using probability distributions of estimates of the likelihood of occurrence and the magnitude of losses. If this is done, then the calculation of risk is not simplistic, but requires the use of simulation techniques to obtain random samples from probability distributions of the independent variables (such as the likelihood of occurrence and magnitude of outcome) in order to arrive at the probability distribution of the dependent variable (i.e., risk). It should be noted, however, that the use of such techniques, the quality of the inputs, and the interpretation of the results all require a level of sophistication higher than that needed for the more simple methods.

Not only must the analyst take a mathematical approach that is more complicated, but he or she must also account for the dynamics of the situation. When it comes to security and safety, the threats, exploits, and vulnerabilities are in a continual state of flux. Some factors are the result of changing technologies and technical environments that facilitate access to software systems over public networks, and enable those with a technical bent, who are not trained in security and safety and are not answerable to any corporate policies and standards, to develop heavily-used apps. Technology advances also allow systems with different security and safety criteria to be combined into systems of systems and cyber-physical systems. Other influences include changing laws and regulations, particularly as they apply to personal data protection and the safety of products. Still other factors are inward looking, where organizations incorporate secure and safe systems development techniques to varying degrees and respond differently to the build-security-in movement.

The dynamics introduced by these factors must be accounted for in the risk calculations. Current and anticipated changes need to be included in risk models, and calculations need to be rerun whenever there are material changes in any of the variables contained in the risk model, or if the decision-makers believe that there will be major new developments that will affect the outcomes. The changes might also suggest that the model itself needs to be revised, as it may be discovered that it no longer accurately represents the environment being considered. Preferably, risk models will be run for a variety or range of scenarios, considering different probabilities of events, different losses or gains, and the like, resulting from new assumptions about known advances in technologies, changes in marketplaces including greater competition, and cheaper sources of product. In summary, it is not adequate to run one risk model or set (✱) of models and then sit back and operate based on the results of those models. Technical innovations, changes in markets, social changes, competing products, staff defections, and the like, all call into question the original results, and so the models need to be rerun to account for these changes. The result might be retrenchment, expansion, changes to the business model, new alliances, and the like. However, it is far better to be aware of these changes and respond to them than to be the victim of obsolete technology and miss business opportunities.

Furthermore, the inputs to the risk models and the interpretation of results are often influenced by personal views. Even when risk assessors are trying to be as objective as possible, personal bias will creep in. For instance, if a professional safety engineer is estimating the risks relating to a control system for which he or she has no direct responsibility, the engineer will want to protect his or her reputation; this is because one's reputation affects one's opinion of one's self, and will also affect future work opportunities. Consequently, there is a need to adjust estimates for personal bias, which is a task that few want to take on because that would mean assuming some of the responsibility and liability in the event that something bad were to occur. The extreme example of this is the tragedy of the commons, wherein no one is responsible or accountable for the risks, although ultimately someone or some group will pay the price if a disaster occurs.

Finally, the view of risk needs to be holistic. The points of view of a broad range of stakeholders need to be considered. It is often the case that different stakeholders, such as vendors, law enforcement officers, regulators, auditors, customers, and users, all have different interests in the security and safety of software-intensive systems. While it is a daunting task to try to elicit such views, it is even harder to incorporate them into risk calculations. This is because one must determine the relative weights to apply to the opinions of each group and come up with a way in which to aggregate those views, which are always changing based upon events and experiences.

In conclusion, if one agrees that improved risk assessment and management is needed, it is necessary to consider and incorporate all the approaches, biases, and responsibilities into one's analysis of risks as they relate to security-critical and safety-critical software systems. If any one of the above factors is ignored, then the risk analyses and the actions they suggest will be flawed, which only serves to increase the risk further. However, it is better to identify risks and perform as good an analysis as one can, even if it is deficient, than to throw up one's hands and abandon risk analysis altogether.

Endnotes

[1]   Gilb, T., *Principles of Software Engineering Management*, Reading, MA: Addison-Wesley, 1988.

[2]   Boehm, B.W., *Tutorial: Software Risk Management*, Washington, DC: IEEE Computer Society Press, 1989.

[3]   Boehm, B.W., "A Spiral Model of Software Development and Enhancement," *Computer*, Vol. 21, No. 5, May 1988, pp. 61–72.

[4]   See http://greenspun.com/bboard/q-and-a-fetch-msg.tcl?msg_id=002quF, last accessed on July 18, 2012.

[5]   See http://www.csoonline.com/article/569263/there-s-an-insecure-app-for-that, last accessed on July 18, 2012.

[6]   See http://www22.verizon.com/Support/Residential/homecontrol/home+monitoring+and+control/overview/129406.htm, last accessed on July 18, 2012.

[7]   See blog at www.threatanalysis.com/blog/?p=43, last accessed on March 23, 2012.

[8]   Parker, D., "Making the Case for Replacing Risk-Based Security" in *Enterprise Information Security and Privacy*, C. W. Axelrod, J. L. Bayuk and D. Schutzer, (eds.), Norwood, MA: Artech House, 2009.

[9]   Borg, S., "The Economics of Loss" in *Enterprise Information Security and Privacy*, C.W. Axelrod, J.L. Bayuk, and D. Schutzer (eds.), Norwood, MA:Artech House, 2009.

[10]   See J. Lowder, "Why the "Risk = Threats x Vulnerabilities x Impact" Formula is Mathematical Nonsense" at http://www.bloginfosec.com/2010/08/23/why-the-risk-threats-x-vulnerabilities-x-impact-formula-is-mathematical-nonsense/ for Part 1, and http://www.bloginfosec.com/2010/08/31/why-the-%e2%80%9crisk-threat-x-vulnerability-x-impact%e2%80%9d-formula-is-mathematical-nonsense-part-2/ for Part 2, last accessed on July 18, 2012.

[11]   Howard, M,. and S. Lipner., *Best Practices: The Security Development Lifecycle*, Redmond, WA: Microsoft Press, 2006.

[12] Axelrod, C. W., "Investing in Software Resiliency," *STSC CrossTalk: The Journal of Defense Software Engineering* (September/October 2009), pp. 20–25. Available at http://www.crosstalkonline.org/storage/issue-archives/2009/200909/200909-Axelrod.pdf. Accessed July 18, 2012.

[13] McGraw, G., *Software Security: Building Security In*, Boston: Addison-Wesley, 2006.

[14] Swiderski, F., *Threat Modeling*, Redmond, WA: Microsoft Press, 2004.

[15] See https://www.owasp.org/index.php/Threat_Risk_Modeling.

[16] See D. E. Sanger, "Obama sped up wave of cyberattacks against Iran," *The New York Times*, 2012. Available at http://www.nytimes.com/2012/06/01/world/middleeast/obama-ordered-wave-of-cyberattacks-against-iran.html?_r=1&ref=davidesanger, last accessed on July 18, 2012.

[17] Tables 6.2 and 6.3 were derived from D. Blondin, "Would certification become mandatory in automotive engineering?" *ERTS Second European Congress*, 2004. Available at http://www.axlog.fr/R_d/documents/blondin2004.pdf , last accessed on July 18, 2012.

[18] See http://en.wikipedia.org/wiki/Safety_Integrity_Level, last accessed on July 18, 2012.

[19] McConnell, S., *Code Complete: A Practical Handbook of Software Construction, Second Edition*, Redmond, WA: Microsoft Press, 2004. Available online at http://jampad.net/Library/codecomplete/, last accessed on July 18, 2012.

[20] See L. Dignan, "Global Payments financial hit over breach likely manageable," available at http://www.zdnet.com/blog/btl/global-payments-financial-hit-over-breach-likely-manageable/72897?tag=mncol;txt, last accessed on July 18, 2012.

[21] See Belady [32]. Last accessed on July 18, 2012 at http://cseweb.ucsd.edu/~wgg/CSE218/BeladyModel-10.1.1.86.9200.pdf.

[22] See M. Specter, "Annals of Medicine: The Deadliest Virus," *The New Yorker*, March 12, 2012, available at http://www.newyorker.com/reporting/2012/03/12/120312fa_fact_specter, last accessed on July 18, 2012.

[23] See http://en.wikipedia.org/wiki/Morris_worm, last accessed on July 18, 2012.

[24] See E. Felton, "New Research Result: Cold Boot Attacks on Disk Encryption," Freedom to Tinker blog, February 21, 2008, at https://freedom-to-tinker.com/blog/felten/new-research-result-cold-boot-attacks-disk-encryption, last accessed on July 18, 2012.

[25] OCC Bulletin 2008-16 on Application Security is available at http://www.occ.gov/news-issuances/bulletins/2008/bulletin-2008-16.html, last accessed on July 18, 2012.

[26] See http://www.bloginfosec.com/2011/09/12/risk-mismanagement-%e2%80%93-scoring-vs-monte-carlo-vs-scoring/, last accessed on July 18, 2012.

[27] Jones, J. A., *An Introduction to Factor Analysis of Information Risk (FAIR)*, Risk Management Insight, 2005. Available at www.riskmanagementinsight.com/media/docs/FAIR_introduction.pdf. Accessed July 18, 2012.

[28] Hubbard, D., *The Failure of Risk Management: Why It's Broken and How to Fix It*, Hoboken, NJ: J. Wiley and Sons, 2009.

[29] Hardin, G., "The Tragedy of the Commons," *Science*, Vol. 162, 1968, pp.1243–1248.

[30] Angwin, J., and J. Singer-Vine, "What They Know: The Selling of *You*," *The Wall Street Journal*, April 7–8, 2012, pp. C1 – C2.

[31] Axelrod, C. W., "The Dynamics of Privacy Risk," *ISACA Information Systems Control Journal*, Vol. 1, 2007. Available at http://www.isaca.org/Journal/Past-Issues/2007/Volume-1/Documents/jpdf0701-the-dynamics-of-privacy.pdf. Accessed April 8, 2012.

[32] Belady, L .A., and M. M. Lehman, "A Model for Large Program Development," *IBM Systems Journal*, No. 3, 1976, pp. 225–252.

7

Software System Security and Safety Metrics

Not everything that can be counted counts, and not everything that counts can be counted.
—Attributed to Albert Einstein (Originated by William Bruce Cameron [1])

If you cannot measure it, you cannot improve it.
—Lord Kelvin [2]

What you measure affects what you do ... if you don't measure the right thing, you don't do the right thing."
—Joseph E. Stiglitz, Nobel Laureate (as quoted in [3])

There is no accepted agreement on software safety measures and metrics.
—Andrew J. Kornecki, Embry Riddle Aeronautical University, [4]

Introduction

The above quotations are clearly contradictory. Lord Kelvin and modern day thought-leaders such as Gary McGraw, Dan Geer, and Andrew Jaquith (to name a few), believe that one cannot improve that which one is not able to measure. However, Albert Einstein and William Bruce Cameron, together with Gary Hinson and others, support the view that there are system attributes, which may not be measurable in numeric terms but which nevertheless can be used to significant advantage in assessing risks. Joseph Stiglitz, quoted above, extends the argument beyond *how* one might measure to *what* one should indeed measure.

131

Hinson [5], Hubbard [6], and Axelrod [7] believe one can use characteristics that are not simple to measure or that may not in fact be considered measurable in the traditional sense for decision-making purposes.

If one can come up with a preferred method for risk identification, assessment, and management (which is no easy task), further questions arise about being reasonably confident that the resulting risk analyses are based on realistic and meaningful metrics. Here is a list of some of those questions:

- What information must be obtained in order to calculate risk using a specific selected method?
- To what extent is historical and real-time data useful, relatively speaking?
- How quickly must data collection, analysis, and reporting be done in order to ensure that decisions can lead to timely actions?
- How accurate does the data have to be for one to be reasonably sure of making the right decisions?
- Is required data readily available for effective decision-making—if not, what has to done to obtain it?
- What level of effort is needed to generate, collect, analyze, report, and respond to specific data?
- Is the additional effort required to obtain the data worth the incremental value of such data?
- To what extent are risk metrics constrained by the availability and suitability of data?

The most valuable security-related and safety-related data about software-intensive systems may well be more difficult and expensive to collect, maintain, and analyze than run-of-the-mill, readily-available data. Furthermore, easily-available data may be less useful for calculating software security and safety risks and for making useful decisions than data that is more difficult and costly to come by. Of course, it is difficult to prove these assertions other than from recognizing that there are deficiencies in current metrics. One would expect that the state of software-system security and safety would be much better if security and safety metrics were more accurate and complete.

It is important to define careful measurements and metrics and note the differences between them. This includes pointing out the differences between data and metrics, measurements and metrics, and so on.

Obtaining Meaningful Data

In the 1970s, IBM used the phrase "Not just data, reality" in their advertising. This slogan is especially insightful when it comes to gathering and analyzing software security and safety data. It is not useful to collect extraneous or irrelevant data, even if collecting such data costs virtually nothing (there will likely still be substantial data management and storage costs to absorb, however), nor is it beneficial to collect particularly pertinent data if there is no way to implement any decisions that might result from it. There is usually a trade-off to consider when determining what to spend on obtaining specific data because there may well be diminishing returns of such data collection, particularly over time. For example, it is much less useful to learn that your systems are vulnerable to a particular exploit after they have been successfully attacked than to obtain such information sufficiently in advance to be able to apply a patch to negate the impact of an attack. Furthermore, one must allow for the cost and time required to convert data into the "reality" of metrics; it is often not enough to collect information quickly—it must be converted into useful actionable information on a timely basis.

Defining Metrics

There are many definitions of the word *metrics*. Here are a few of them:

- "Metrics are tools designed to facilitate decision making and improve performance and accountability through collection, analysis and reporting of relevant ... data" [8].
- "Metrics report how well policies, processes and controls are functioning, and whether or not desired performance outcomes are being achieved ..." [9]
- "A metric is a quantitative measure of the degree to which a system, component or process possesses a given attribute" [10].

A common theme throughout these definitions is that the purpose of metrics is to lead to decisions about how to enforce policies and manage processes and procedures. It is not sufficient that a metric be intellectually interesting. Nor should metrics and analytical results be taken at face value without giving some thought as to their meaning, particularly with respect to cause and effect. There appears to have been a recent proliferation of analyses, particularly in the health and nutrition fields, that discover relationships that do not necessarily have the underpinning of cause and effect. In many of these cases, researchers commonly include a disclaimer that the results do not imply a causal relationship. This raises the question as to what the value of the analysis really is in the first place, as the researchers seem to be suggesting that the public should not respond to the results. This begs the question as to why researchers published the findings at all. Furthermore, it is quite possible in such circumstances that the analytical results will lead to wrong and possibly damaging decisions and actions on the part of decision-makers. A similar situation exists in the system security and safety space in that many metrics are published but they cannot be used to protect against attacks, prevent a system from doing damage, and the like.

Unfortunately many reported metrics relating to information security are not very useful. For example, it is interesting to learn how viruses and other malware are growing exponentially in volume and effectiveness. However, a decision-maker needs to know whether these same viruses, say, will have any impact over the domain for which he or she is responsible. If there is no impact, then those metrics are not actionable and therefore not useful. In fact, they may be detrimental because of the cost of gathering, reporting, and discussing them.

Andrew Jaquith [11] defines a metric as "a consistent standard for measurement." He states that a good metric should be:

- Consistently measured;
- Cheap to gather;
- Expressed as a cardinal number or percentage;
- Expressed using at least one unit of measure.

It was argued earlier that "cheap to gather" is not necessarily a good criterion for metrics, so this requirement is questionable. Also, the use of cardinal and percentage metrics is often misleading because they often contain too little information for useful decision-making. In particular, a numeric metric out of context, such as: "We applied 200 patches last month," provides no sense as to what level of patching is adequate. A percentage metric, such as: "We patched 90 percent of our servers last month," does not indicate how critical the servers, particularly those that were not patched, might be. Frequently, some explanatory wording is needed to make these metrics more meaningful, such as "The 10 percent of servers that were not patched are considered the least critical to secure operations."

As a real-world example of misleading metrics, consider an actual case in which full-disk encryption had been applied to 95 percent of a company's laptop computers, with a goal of encrypting the remaining machines when employees brought them in for service. However, one of the remaining 5 percent, on which the disks were not encrypted, was stolen. This particular laptop computer held personal information of hundreds of thousands of customers. The out-of-pocket costs to the company of this loss exceeded one million dollars, with indirect and less tangible costs possibly doubling that amount. The metric of 95 percent completion might have been considered good in many situations—but not in this one.

Differentiating Between Metrics and Measures

The difference between a measure and a metric is expressed and illustrated in a number of places. For instance, in "Risk Measure and Risk Metric," we read the following [12]:

- A *measure* is an operation for assigning a number to something.
- A *metric* is our interpretation of the assigned number.

As an example, if someone stands on a scale, the process of noting the number of pounds (or kilograms) on the dial is a measure, whereas the interpretation of that number to determine that a person is overweight turns it into a metric that can be used for decision-making, such as deciding to go on a diet.

This is somewhat in line with the differentiation proposed above; the mere obtaining of a number is not a metric—only when a number is interpreted can it be used for decision-making and then decisions can lead to actions. This concept is illustrated in Figure 7.1.

The diagram shows, by means of narrowing arrows, that there are losses as the process progresses. That is to say, not all measurements can be converted into metrics; decision-makers are likely only to use a subset of metrics to make their decisions; and action-takers might only respond to certain decisions and not to others. For example, one might have access to weather forecasts for virtually any location on Earth. However, temperature alone is not useful until it can be attributed to a particular geographic area in which the decision-maker happens to be or to which he or she is planning to travel (although, of course, others will be interested in forecasts for other regions). Based on the temperature forecast, the decision-maker might determine that it is appropriate to wear and/or pack particular sets of clothing and take the actions necessary to find and don that clothing and/or include suitable clothing in his or her luggage. The decision-maker and action-taker may be the same person, as in the weather example. However, in many organizations, those making decisions and taking action are often different individuals or groups, with managers making the decisions and workers implementing those decisions.

Another source that differentiates between measures and metrics is the National Institute of Standards and Technology Software Assurance Metrics

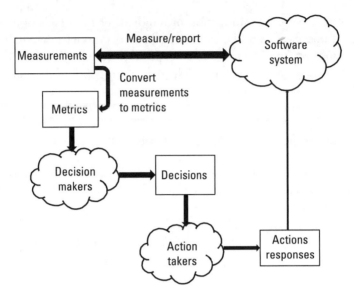

Figure 7.1 From measurement to action.

and Tool Evaluation (NIST SAMATE) Project [13], which presents the following definitions:

- The term *measure* is used for more concrete or objective attributes.
- The term *metric* is used for more abstract, higher-level, or somewhat subjective attributes.

An example of this, provided on the NIST website, is that the number of lines of program code in a computer program is a measure, whereas the robustness, quality, and effectiveness of the code are considered to be metrics. This is somewhat consistent with the *Risk Glossary* definitions [14] if one were to consider the number of pounds shown on a scale as being concrete and objective, and the concept of weight as being more abstract. In fact, the term *weight* is often misapplied; the difference between weight and mass should be taken into account; where "[m]ass is a measurement of the amount of matter something contains" and weight is "the measurement of the pull of gravity on an object." Therefore, *weight* needs to be defined in a specific context, since someone of the same mass will weigh one sixth as much on the moon as they do on Earth due to the moon's lower gravity. This is a significant aspect of software systems, where metrics generally only apply in a particular context, such as operating system, platform, network configuration, and the like.

Another definition of measurement in Munson [15] is:

... the mapping of numbers to attributes of objects in accordance with some prescribed rule ... The mapping must preserve intuitive and empirical observations about the attributes and entities.

Munson is not as specific as NIST in assigning descriptors like *concrete* and *objective* to a measure. However, one might argue that his definition of *measurement* is more general and can apply to both measures and metrics insofar as both objective and subjective attributes (measures and metrics, respectively) are subject to some form of measurement.

The question then arises as to whether qualitative measurement applies to both measures and metrics. After all, according to NIST, the difference is in the objectivity or subjectivity of the attributes, not the means of obtaining them. For an answer to this question, we refer to Cruickshank [16], who states that:

> Empirical observation requires experimentation. Therefore ... for a software metric to be valid, it must be based on empirical observation through experimentation, whether qualitative or quantitative.

Consequently, we should understand that measures and metrics can be either quantitative or qualitative and, in either case, they provide a better picture as to what is going on within a software system than not having the measures or metrics. In Table 7.1, we show examples of the relationships among these

Table 7.1
Examples of Metrics and Measurement Attributes for an Automobile

Attribute	Quantitative Measurement	Qualitative Measurement
Measure (Objective)	The speed of an automobile in miles per hour (mph) as shown on the speedometer.	The relative time that a vehicle takes to reach 60 mph from a standing start (e.g., slow, average, fast).
Measure-related Decision	Is the vehicle traveling within a numeric speed limit, say 55 miles per hour?	Does the car accelerate quickly enough for my needs (and wants)? For example, can I beat my colleague's car from a standing start.
Measure-related Action	If yes, no action required. If not, slow the vehicle down until it is traveling at or below the speed limit.	Select that vehicle based on relative acceleration if that is primary or a highly significant factor.
Metric (Subjective)	The revolutions per minute (rpm) as shown on the tachometer.	The sound of the vehicle's exhaust system.
Metric-related Decision	When should I change gear?	Does the sound produce the desired emotional feelings?
Metric-related Action	Change gear up or down when at a particular revolutions-per-minute limit (say, 2,000 rpm)	Choose a vehicle that provides a desirable feeling with respect to the sound of the exhaust system, or replace existing exhaust system with one that produces the desired sound.

factors with respect to an automobile. The example illustrates the differences between objective and subjective attributes, and whether the numeric value of the attribute needs to be accurate or can be relative.

The same categorization can be used for safety-critical and security-critical software systems, and we will go through a similar exercise for them. Software is less tangible (or more abstract) in many ways than an automobile; it is not as easy to come up with examples within each category. As Cruikshank [16] puts it: "... the abstract nature of software can often preclude quantitative measures of certain quality attributes." Because safety and security are essentially quality attributes, the challenge is therefore to come up with quantitative, but meaningful, measures so that very specific decisions can be made. This is the challenge that we attempt to address in this book from both the demand side (i.e., determining which measures are needed) and the supply side (i.e., creation of requisite measures and metrics).

In Table 7.2, we look at the uses of measurements and metrics to determine how useful they are for the purposes of deciding what actions to take. In general, measurements that have not been processed into metrics are not particularly useful for decision-making. However, some metrics also lack appropriateness for decision-making, which is the main criterion against which they should be judged.

Software Metrics

One definition of software metric is "... a measure of some property of a piece of software or its specifications" [17]. This definition differs in a fundamental

Table 7.2
Comparing Measurements and Metrics with Respect to Use and Usefulness

Use	Measurements	Metrics	Comments
General usefulness	Measurements lack context and do not have the attributes necessary to foster decision-making and action.	Some metrics are insufficient for appropriate decision-making – need to include context and other factors such as value and uncertainly.	Concern about the overly subjective interpretation of metrics.
Use for decision-making	Measurements are often not particularly useful for decision-making.	Some metrics do support effective decision-making, but others can be very misleading.	Often metrics that are the most useful for decision-making are those that are more costly and difficult to collect.
Actionable	Usually, actions cannot be derived directly from measurements.	In many cases, metrics can be the basis for decisions and actions; sometimes not.	The ability to take effective action depends on many factors, some of which may not be under the control of either those taking the measurements or decision-makers.

way from the metrics above. In fact, this definition does not require that metrics are input for decision-making and that the decisions should be actionable. If not actionable, then such a "property becomes a measure, not a metric. On the other hand, all metrics are not actionable according to the common view, although it would be preferable if they did lead to positive action.

Cruickshank [16] draws the following conclusions, which includes support for nonnumeric metrics with respect to software metrics:

> Software metrics are ... quantitative measurements of ... product (system or component), process, or even project ... indicating the quality of a desired attribute. However, software metrics can be concerned with more than just quantitative measurements. Since we are measuring quality of product, process or project, qualitative aspects must be considered. *Metrics can also be qualitative in nature* [emphasis added].

On the other hand, McGraw [18] favors numerical representations of measures and metrics, where "... [m]easures are numeric values assigned to a given artifact, software product, or process" and "... [a] metric is a combination of two or more measures that together provide some business relevant meaning." McGraw provides an example of a metric, namely, "the number of breaches per lines of code," which represents a "security defect density," as opposed to "the number of breaches," which he defines as a *measure*. McGraw [18] echoes Cruikshank's assertion [16] that metrics can be obtained for project, process, and product, and adds a fourth area—namely, organization. He also points out that processes and controls must be put in place early in the development life cycle in order to provide the requisite metrics, which is a concept that is applied in subsequent chapters, although we will be address the approach more in terms of the application (or product) generating metrics rather than looking to the processes and controls.

It is worth noting the tendency of security and safety professionals to take a very specific view of their areas of responsibility, combined with an underlying need to quantify metrics. These characteristics make for a strong bias towards quantitative measures, that is, the numeric values of concrete, objective measures. The problem is that software-intensive systems are much less amenable to the application of precise measures. This means that the exercise to convert software measures to metrics for decision-making and action is often lacking. As a result, the wrong decisions are often made and inappropriate actions taken based on insufficient and misleading metrics. Because this focus on quantitative measurement and very specific decisions is part of the culture of an engineering background, it will likely take considerable time before researchers and practitioners evolve a broader view of software systems and define a full range of requirement to be met.

Measuring and Reporting Metrics

When it comes to measuring security-related and safety-related metrics, there are a number of approaches with varying degrees of merit. A specific taxonomy, as suggested in Axelrod [7], is shown in Table 7.3.

The advantages and disadvantages of the above categories of metric are shown in Table 7.4. The weighting of such metrics is highly dependent upon the nature of the organization from which the data is being collected, as well as who is collecting and analyzing the data. Often, an outside consultant will have more success in obtaining useful data and performing objective analyses.

In Figure 7.2, we show the behavior of cost and value estimates as the complexity of the metrics approaches increases. In general, as the data collection and analysis becomes more complex, the costs will increase and it is expected that the value, as it relates to decision-making, will increase even more. Although the diagram shows value growth increasing relative to the costs of increasingly complex metrics and risk approaches, value might not necessarily keep up with costs as the data collection and risk models become more sophisticated. In part, this can be due to the benefits being less tangible when value and uncertainty are taken into account.

Figure 7.2 indicates a range of values and costs with upper and lower bounds. Because of the ranges shown, the curves suggest that, at any level of complexity, the net value of the exercise, namely the usefulness of the metrics category for decision-making less the cost of data collection and analysis, can in fact be negative. As costs increase along with the complexity of the approach, so the value of the results also increases. It is up to the risk manager and/or analyst to arrive at an optimal approach with respect to net value. The optimal approach might lie in the range of the holistic, value, and uncertainty approaches because they produce the most meaningful representations of risk.

Table 7.3
Taxonomy of Categories of Metrics

Category of Metric	Description of Responses
Existence	"Yes," "No," or "Not Applicable (N/A)"
Ordinal	Typically "High," "Moderate," or "Low"
Score	On a scale, such as "one to ten"
Cardinal	A number in response to the question: "How many…?"
Percentage	A number in response to the question: "What percentage of…?"
Holistic	More complete view from authoritative source
Value	Typically, total net loss in value for different approaches
Uncertainty	Stochastic or probabilistic approach

Table 7.4
Advantages and Disadvantages of Various Categories of Metric

Category of Metric	Advantages	Disadvantages
Existence (Yes, No, N/A)	Easy to collect Does not require much thought at all on the part of responder Provides a first cut to help determine if certain areas should be analyzed in greater detail	Provides latitude for the responder to misinterpret questions, whether accidentally or deliberately, and provide incorrect answers "No" answer indicates a problem, but "Yes" and "N/A" do not ensure compliance May not account for context
Ordinal (Low, Moderate, High)	Easy to collect and further analyze the data Does not require very much thought on the part of responder Provides a high-level assessment and might point to issues that should be pursued in greater detail	Answers are highly subjective and cannot be readily compared across responders Easy for responder to hide problems by responding that there is a low risk, say, when the responder might really think otherwise, in large part because of the highly subjective nature of the responses May not account for context
Score (e.g., 1 through 10)	Moderately easy for responder to assign numbers Relatively easy to collect and analyze the data Allows for increased granularity relative to the above metrics	Requires responder to try to determine the scale Answers are likely to be highly subjective with a strong bias in favor of the responder Easy for responder to hide problems by arbitrarily assigning numbers Summation and averaging of results can be misleading May not account for context
Cardinal (How many ...?)	Usually quite easy to collect data Can be very easy to analyze Allows for practically any level of granularity	May not know the size of total population May not necessarily account for the context Analyst does not usually know how significant the measured population is
Percentage (What percentage ...?)	Moderately easy to collect Can be very easy to analyze Can account for various levels of granularity Measured population is known	Analyst does not know significance (e.g., quality) of measured population versus total or other populations May not necessarily account for context
Holistic (All-encompassing)	Represents a more realistic and relatively complete list of risks to evaluate Involves broader range of respondents and decision-makers More likely to consider and account for context	Difficult to define complete set of measures to collect Difficult to collect objective data because respondents may cut across organizational lines Usually difficult to compare metrics and aggregate them

Table 7.4 (continued)

Category of Metric	Advantages	Disadvantages
Value (Loss in value)	Value, particularly net value (value less cost), is a more realistic representation of residual risk than obtained from other methods Involves a broader range of perspectives from respondents and decision-makers	Due to the highly-subjective nature of value estimates, values may not be easily compared and aggregated Difficult to define complete set of measures to collect Difficult to collect objective data because respondents may cut across organizational lines
Uncertainty (Probability distributions)	Representing uncertainty is much more realistic than using point estimates Requires respondents to think more deeply about consequences of decisions	Requires respondents to understand probability distributions and basic statistical methods Due to the highly-subjective nature of probability estimates, results may be difficult tp compare and aggregate—will generally need to use a computer model (as opposed to a spreadsheet) for calculations Difficult to collect objective data because respondents may cut across organizational lines

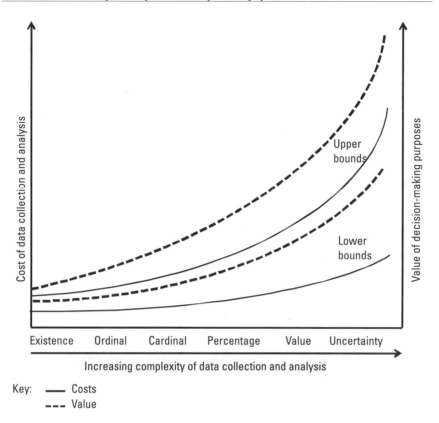

Figure 7.2 Cost versus value of various metrics approaches.

In Figure 7.2, the ranges of the cost and value curves are shown to increase generally with complexity, denoting the higher level of uncertainty that accompanies the more complex methods. There is also considerable uncertainty when it comes to the quality of the data gathering and analysis processes as approaches become more sophisticated. Clearly, it does not take the same level of expertise to collect and analyze ordinal data as it does to collect and analyze stochastic or probability-related data. As Hubbard [6] correctly points out, respondents need to be trained in how to assess probabilistic situations and analysts must have a background in the pragmatic use of statistics.

Metrics for Meeting Requirements

Requirements can be classified as *functional* or *nonfunctional*; those terms are defined as follows, per Braude [19]:

- *Functional requirements* (or *behavioral requirements*) specify services that the application must provide.
- *Nonfunctional requirements* include all requirements not considered to be in the class of functional requirements.

Of course, many have pointed out that the latter term could be interpreted to mean "requirements that do not function," but that is not how it is commonly meant in systems engineering. The usual categorization of attributes, such as safety and security, as nonfunctional requirements results in the lack of attention that is usually accorded these requirements in any but the most mission-critical of systems.

 A major point of this book, which is often given short shrift in other texts, is that many so-called nonfunctional requirements also have their functional counterparts. The separation of knowledge and duties is such that those with expertise and responsibility in the functionality of software systems are generally not knowledgeable about, or even interested in, the nonfunctional attributes of those same systems. Many experts in the nonfunctional aspects of software systems, many of whom do not know or care about the detailed functionality of the systems. This state of affairs might be acceptable were it not for the fact that functional aspects of software systems have considerable impact on nonfunctional requirements and vice versa.

Table 7.5 describes the nonfunctional attributes, as mentioned in Braude [19], and some that are in MISRA [20]. In particular, the omission of safety and interoperability as significant attributes is typical of books, papers, and presentations that focus on general software engineering, such as Braude [19] and security engineering, such as Anderson [21]. Safety, in particular, is usually only included in those publications specializing in that attribute, and these publications are likely to omit consideration of security, as in Herrmann [22].

Similarly, there is usually little reference to relationships between software functionality and nonfunctional requirements in security-oriented publications, whereas safety attributes are considered to be tightly interrelated with software functionality. In Table 7.6, we show some of the more significant nonfunctional requirements that impact the security and safety of various software functions.

Sometimes certain nonfunctional attributes, such as security-related authorization or encryption, are written into the functionality of the applications. While commonly done, such a practice is suboptimal and can lead to errors and inappropriate use, as opposed to when such nonfunctional services are isolated from functional applications.

There are usually serious issues regarding organizational structure and the expertise and biases of those with responsibility for determining requirements and ensuring that they are included and enforced throughout the secure

Table 7.5
Descriptions and Uses of Nonfunctional Attributes

Nonfunctional Attributes	Description/Use
Reliability*	Determines availability and integrity of systems and data
Availability*	Determines whether or not systems can be used for their intended purposes when they are needed
Performance*	Includes speed, throughput, storage capacity. Describes responsiveness of system and timeliness of its functional services—lack of performance leads to lower productivity, loss of users, higher operational costs, etc.
Security*	Protection of information assets from incidents resulting in data breaches, loss of service, fraud, etc. Components stated as CIA: confidentiality, integrity, availability, plus others as listed below.
Confidentiality*	Data not accessible by unauthorized subjects—included in security requirement
Integrity*	Included in security requirement Also a safety requirement (malfunction)
Availability and reliability*	Affects average uptime—availability included in security requirement Also a safety requirement (failure)
Nonrepudiation*	Proof of existence of agreements—included in security requirement
Authentication*	Ability to validate users' identities—included in security requirement
Authorization*	Permission to deal with subject—included in security requirement
Maintainability*	Increases availability and reduces malfunctioning and failure rates
Portability*	Ability to run software on different platforms and infrastructures with minimal or no modification
Error handling*	Speed and correctness of error resolution, error routines, self-healing systems, fail safe—affects availability
Interoperability	Ability to access data across applications running on different platforms—affects availability
Resiliency	Combination of error handling, redundant systems, responsiveness—affects availability, and affected by portability and interoperability of applications
Safety[†]	Affects damage to external parties or environment from software system malfunctions or failure
Complexity[†]	Increased complexity affects likelihood that system will contain errors
Maintainability[†]	See above
Modularity[†]	Higher use of modules implies less complexity and therefore lower likelihood of errors
Structure[†]	Good structure with respect to control, data and information flow improves likelihood of reduced errors
Testability[†]	Ease with which software systems can be tested with a high level of confidence that errors will be detected
Reliability[†]	See above

* *From:* Braude [19].
[†] *From:* MISRA [20].

software development life cycle (SSDLC). This will be further addressed later when we discuss the development process. Here, we merely draw attention to many of the nonfunctional attributes that exist and to the unrealistic hope that the same individuals who are fluent in the functionality of software might also understand the full implications of all the nonfunctional attributes listed in Table 7.5. It may even be optimistic to believe that applications developers understand and care about even a couple of the nonfunctional attributes. It is little wonder that nonfunctional requirements do not appear in software system specifications.

Risk Metrics

There is surprisingly little agreement as to which metrics should be used. That is particularly true of risk-related metrics for security-critical and safety-critical systems. It should be noted here that there does not seem to be anywhere near the same level of confusion and debate when it comes to metrics relating to the operation of safety-critical systems, probably because the latter are more ingrained into the physical world, where metrics are often easier to come by and to interpret than for the more abstract world of software. Nevertheless, Cruickshank [16] points out that metrics relating to the development process for safety-critical software systems are also lacking, particularly with respect to the validation process. These differences are shown in Table 7.7.

Consideration of Individual Metrics

We first look into individual metrics, which are particularly relevant to the security and safety of software systems, and then consider the superset of metrics

Table 7.6
Typical Nonfunctional Requirements for Various Software Functions

Software Function	Nonfunctional Requirements Affecting Software Function
Read	Authentication, authorization, integrity, confidentiality
Write	Authentication, authorization, integrity, confidentiality, nonrepudiation
Modify	Authentication, authorization, integrity, confidentiality, nonrepudiation
Delete	Authentication, authorization, nonrepudiation
Save	Authentication, authorization, integrity, confidentiality, nonrepudiation
Transfer	Authentication, authorization, integrity, performance, confidentiality, availability
Calculate	Authentication, authorization, integrity, performance, availability
Display	Authentication, authorization, integrity, confidentiality, availability
Print	Authentication, authorization, integrity, availability, performance

Table 7.7
Understanding of Risk Metrics for Security-Critical and Safety-Critical Software Systems

Characteristics	Security-Critical Software Systems	Safety-Critical Software Systems
Attributes considered	Risks are mostly considered for nonfunctional attributes and likelihood that the systems will be compromised by an attacker	Risks are mostly related to the dangers of malfunctioning and failure
Risk evaluation and testing	Security risk evaluation and testing relates mostly to platforms and networks, with relatively little attention given to the functional security aspects of software	Verification of safety standards conformance is rigorous for critical systems and involves the functional aspects of systems and the level of hazard that they would represent if they were to malfunction or fail

applicable to software-intensive systems of systems or cyber-physical systems, which must meet both security and safety requirements.

Historically, these two areas have not been related to one another. For example, if you read Merkow [23] on the development of secure and resilient software, and Hermann's book [22] on the safety and reliability of software, you will see that there is no mention of safety in [23], and no mention of security in [20]. It is also interesting to note that books specializing in security metrics usually pay little attention to application or software system security metrics. For example, Jaquith [11] devotes a mere dozen pages to the topic.

We can further support this lack of focus on application software security metrics by looking at popular works on the development of secure software and see the extent to which the authors discuss metrics. For example, McGraw [18] covers the topic of establishing a metrics program in a single page, and Howard and Lipner [24] omit measures or metrics completely.

Generating requisite metrics is crucial in order to monitor and report questionable activities within applications, and hence be assured of the effectiveness of their security and safety methods. Knowing what is going on within an application is crucial to ensuring the secure and safe operation of mission-critical software-intensive systems. The apparent lack of interest in the topic by thought leaders in the software security area is disconcerting. In fact, the disconnect between software development teams and those requiring additional instrumentation could well be a significant contributor to our seeming inability to build highly secure software-intensive systems.

Safety-critical software systems, on the other hand, are not usually subject to the same levels of attack as security-critical systems because their failure rates appear to be very low. Also, looking at press coverage, safety-related software malfunctions and failures appear to be infrequent relative to breaches and failures of security-critical software systems. Safety metrics, as they relate to software systems, are very different from security metrics. This is likely due to the different perspective of those responsible for safety and the restrictions placed upon them in obtaining certifications for their systems.

Table 7.8 provides examples of functional and nonfunctional metrics as they apply to security-critical and safety-critical software systems.

Over the past couple of years, researchers have begun to see the importance of looking at software systems from both the perspective of safety and security at the same time, as in Gutgarts [25]. However, there is relatively little published about the confluence of security and safety from a requirements and metrics perspective. We shall first look at the security and safety requirements of software systems separately, as illustrated in Table 7.9, and then take a combined view.

Table 7.8

Examples of Metrics Relating to Security and Safety

Metrics Category	Security-Critical Software Systems	Safety-Critical Software Systems
Functional	Determination of secure coding deficiencies Monitoring of user activities within applications	Requirement for resistance to malfunction and failure Determination of limits on potential damage incurred
Nonfunctional	Security Dependability Performance Resiliency Availability Quality	Safety Performance Resiliency Availability Quality

Table 7.9

Primary Focus of Software Security and Safety Engineers

	Security-Critical Software Systems	Safety-Critical Software Systems
Software security engineers	Most of the focus is on nonfunctional system attributes Relatively little attention paid to functional security attributes as shown by the minimal discussion in the literature, see Axelrod [26] Relatively little attention paid to context	Limited knowledge or experience in the safety area Generally a lack of attention to safety aspects
Software safety engineers	Limited knowledge or experience in the security area Generally a lack of attention to security aspects	Strong focus on safety attributes as required by various certifying bodies Both functional and nonfunctional attributes are usually considered Some degree of attention paid to context

Table 7.10
Ordinal Metrics for Information and Control Systems

Metrics	Application 1	Application 2	Application 3	Application N
Predominant system type	Information system	Information system	Control system	Control system
Authentication method	Strong	Weak	Strong	Weak
Authentication process controls	Effective	Somewhat effective	Somewhat effective	Ineffective
Authorization method	Formal	Informal	Formal	Ad hoc
Authorization process controls	Somewhat effective	Ineffective	Effective	Nonexistent
Identity and Access Management (IAM) system	COTS product	None	Home-grown system	None
Classification of data	Formal	None	Informal	None
Restrictions on data flows	Highly restrictive	None	None	Highly restrictive
Value of data	Extremely high	High	Moderate	Low
Degree of encryption (in motion, at rest, in storage)	In motion only	In motion and at rest	In motion only	None
Knowledge of secure coding (training)	Moderate	Moderate	Moderate	None
Implementation of secure coding	Moderate	Minimal	High	None
Testing of secure design and coding	Moderate	Minimal	Minimal	None
Completeness of security/ safety data generation and collection	Fairly complete	Minimal	Minimal	Minimal to none
Effectiveness of analysis and reporting of security/ safety data	Moderate	Minimal	High	Minimal
Responsiveness to breaches and malfunctions/failures	Highly responsive	Minimally responsive	Somewhat responsive	Unresponsive
Software complexity	Very complex	Fairly complex	Relatively simple	Fairly complex
Complexity of software-intensive systems	Very complex	Simple	Moderately complex	Highly complex
Context/environment	Integrated system of systems	Public facing	Isolated, internal	Public facing
Accessibility of system by unauthorized persons or systems	Highly restricted	Somewhat restricted	Unrestricted	Unrestricted
Security/safety aspects of platform	High security Low safety	Low security Moderate safety	Low security Low safety	Minimal security Moderate safety
Security/safety aspects of infrastructure	High security Low safety	Low security High safety	Moderate security High safety	Low security Low safety
Current phase of SDLC	Production	Production	Testing	Coding
Overall security posture	High	Moderate	Moderate	Low
Overall safety posture	Low	Moderate	High	Moderate

Security Metrics for Software Systems

There has been a great deal published on security metrics and the processes, tools and measurements for increasing application security. However, most of the focus has been on nonfunctional attributes, which may or may not be independent of the functionality of the application software.

To the extent that application software security metrics do exist, they usually fall into the cardinal or percentage metrics categories described above. Information security professionals appear to be very fond of numbers, such as the number of unsuccessful attacks per week, the percentage of systems that have had the latest patches applied to them, and the like. This is all well and good, except that these numbers frequently require a footnote to explain them. For example, as discussed previously in regard to the encryption of laptop computers, the percentage of servers that have had patches applied often refers to all servers equally. However, the most important aspect of such a metric is whether or not patches have been applied to all critical servers, where "critical" generally means that the servers support mission-critical applications that are essential because they are needed for business continuance, are required for legal and regulatory compliance, and the like. Application security metrics seldom consider value and uncertainty and are consequently of limited value in their effectiveness.

Another deficiency in application security metrics is that they often do not consider the context in which the software is running. This is a serious omission because applications will display different sets of vulnerabilities depending upon the operating systems on which they are based, for example. It also matters very much, from a security perspective, whether an application is Web facing, and from security and safety perspectives, whether the application runs on an isolated internal network with limited access.

We illustrate this issue in Table 7.10 by showing the levels (i.e., ordinal values) for a given set of characteristics of several security-critical information systems and safety-critical control systems. It should be noted that this table is for illustrative purposes and that, in order to actually use the approach in practice, the analyst needs to obtain much more detail about each element in the matrix, preferably expressing them in value and uncertainty (probability distribution) terms, rather than the ordinal metrics shown here.

It is an interesting and useful exercise to consider the implications of each metrics attribute separately and then in combination. For example, applications 1 and 2 are already in production so that any changes at this stage would be extremely costly. However, because the value of the information is classified as very high and high for applications 1 and 2, respectively, it might still be cost-effective to go back and correct certain security and safety deficiencies. It is doubtful that a major rework could be justified.

Application 3 is not yet in production so that the results from the testing can still be incorporated into the system at a relatively moderate cost. Application N is only at the coding stage so that it is comparatively inexpensive, and not likely to delay implementation much, to modify the programs, although, if a major design change were recommended, it could cause much more significant increases in cost and effort and could delay implementation. On the other hand, the value of the information processed by application N is comparatively low, so that it would likely not make sense to invest much in making the system more secure. However, it should be noted that even if a control system does not contain valuable information as such, malfunctioning and failures could have a huge impact. That is to say, unauthorized access to the application could lead to an outside force taking over control of the system and forcing the system to perform inappropriate and dangerous tasks. In this case, protection has more to do with bad consequences than it does to exposure of sensitive personal data and intellectual property. Application 3 is shown as a control system running on a low-safety platform, yet running on an infrastructure rated high for security. Whether this is acceptable or not will depend on a combination of factors.

The above indicators bring up issues regarding how to weight each attribute and how to aggregate them. Clearly, the interaction of the various attributes is complex and one's ability to come up with precise metrics is very limited. Nevertheless, the exercise itself has considerable value because it will highlight those important characteristics that may be lacking and help focus on which defects and deficiencies should be fixed first, or at all.

Incorporating numeric values into the above evaluation might appear to add precision to the exercise, but in most cases it will not. In fact, there is a considerable danger that using numeric values will be counterproductive. We might ask, therefore, how one might incorporate value measures and uncertainty into a situation that is already fraught with issues with respect to the true representation of subjective ordinal measures of the attributes. The answer is that, if one were able to develop, as an example, a probability distribution of the losses that might be incurred as a result of an inadequate infrastructure or of poor security coding, then one would get a more useful picture of the risk involved in not remediating certain deficiencies.

Safety Metrics for Software Systems

Metrics related to safety-critical software-intensive systems are most concerned with the impact of a malfunction or failure of the system. The hardware components of such systems are always subject to failure, but there are well-established approaches to dealing with them, usually by swapping out components or the like. That is to say, the impact of a hardware failure is usually relatively easy to identify and evaluate, and the response is generally quite straightforward. The exception to this is if it is suspected or shown that the equipment has a weakness or has been compromised in some way; it then serves no purpose to replace the hardware component with a new one that is also defective. The same goes for software.

The root causes of software malfunctions and failures are often less easily identified than are hardware failures. Sometimes a hardware failure is caused by a software error, or vice versa, which can confuse the determination of the reason for the failure or malfunction. For example, if a telecommunications network goes down, the halt in traffic may be because of an equipment failure, because of software errors, or due to human error.

Consider, for example, the actual case where a particular trading desk telephone system would fail and reboot repeatedly. It took engineers weeks to determine the problem. Apparently, in the pristine laboratory environment, there was no noise from the cables, which had short runs and were carefully installed. In the field, with less-than-perfect cabling, the software control system kept invoking a recovery routine that had never been tested in the lab because the carefully-installed cables never malfunctioned. It was determined that there was a coding error in the recovery routine that was activated by signal losses and noise on the cables, and caused the entire system to fail.

As another example, there has generally been a fair amount written recently about securing industrial control systems that are increasingly being exposed to public networks. Numerous articles, congressional hearings, and the like, about securing the smart grid fall into this category. However, there is still relatively little published on trying to reconcile the major conceptual differences that exist between safety and security software engineers' views on system requirements, which is the basic deficiency in the development processes for these systems of systems and cyber-physical systems.

Summary and Conclusions

The development and use of security and safety metrics, as they apply to software systems, are definitely works in progress. Current approaches do not generate all of the requisite measures and metrics, and those that are produced are often insufficient to make appropriate decisions and take necessary actions. As a result, organizations are frequently unaware that successful attacks and breaches have taken place and, when they are notified that such an attack or breach has occurred, they are usually at a loss in determining what happened, who did it, and what to do about it.

The first recommendation for getting a handle on this problem is to understand the range of metrics categories that are available and used, and then to arrange to create, collect, and analyze the data needed to produce meaningful metrics. The resulting risk analyses can be used to determine what actions need to be taken in preparation for and in the wake of an attack or data breach.

This is not a trivial effort, as the most useful metrics for risk identification, assessment, and management are often more difficult to collect and interpret than metrics of lesser value. Post-processing the metrics in order to arrive at aggregated values is also a challenge. Often, respondents need to be trained on how to answer surveys, and analysts need to understand the consequences of threats, exploits, and vulnerabilities and be able to produce results and recommendations in sufficient time to take mitigation actions

It also seems that metrics appropriate for security-critical software systems are not particularly useful for safety-critical software systems, and vice versa. Yet there is a need to exchange methods and procedures between security and safety silos, particularly as the trend is to combine security-critical and safety-critical systems into complex systems of systems and cyber-physical systems, which must demonstrate the ability to maintain a strong security posture and, at the same time, exhibit safety characteristics.

(✱) The basic recommendation here is to understand the strengths and weaknesses of various types of measurements and metrics so that one can determine the best data to collect in order to perform risk analyses that will lead to appropriate mitigation.

Endnotes

[1]　See http://quoteinvestigator.com/2010/05/26/everything-counts-einstein/, last accessed on July 20, 2012.

[2]　See http://zapatopi.net/kelvin/quotes, last accessed on July 20, 2012.

[3]　Goodman, P. S., "Emphasis on Growth is Called Misguided," *The New York Times,* 2009. Available at http://query.nytimes.com/gst/fullpage.html?res=9B02EED7143EF930A157 5AC0A96F9C8B63, last accessed on July 20, 2012.

[4]　Kornecki, A. J., "Assessment of Safety Software via Catastrophic Events Coverage." Available at http://www.google.com/url?sa=t&rct=j&q=kornecki%20safety%20catastrophic&source=web&cd=3&ved=0CDMQFjAC&url=http%3A%2F%2Fciteseerx.ist. psu.edu%2Fviewdoc%2Fdownload%3Fdoi%3D10.1.1.13.457%26rep%3Drep1%26ty pe%3Dpdf&ei=Vt6VT6OiE4ig6QGgyKiPBA&usg=AFQjCNHBxxKa96LDaM8XlsF 7w_GYL7gwgQ, last accessed on July 20, 2012. {

[5]　Hinson, G., "Seven Myths about Information Security Metrics," *ISSA Journal,* Vol. 4, No. 7, July 2006.

[6]　Hubbard, D. W., *How to Measure Anything: Finding the Value of Intangibles in Business, Second Edition,* Hoboken, NJ: John Wiley and Sons, 2010.

[7]   Axelrod, C. W., "Accounting for Value and Uncertainty in Security Metrics," *ISACA Journal*, Vol. 6, 2008. Available at http://www.isaca.org/Journal/Past-Issues/2008/Volume-6/Pages/Accounting-for-Value-and-Uncertainty-in-Security-Metrics1.aspx. Accessed July 19, 2012.

[8]   Swanson, M., et al, *Security Metrics Guide for Information Technology Systems,* NIST Special Publication 800-55, National Institute of Standards and Technology, July 2005, p. 9.

[9]   Government Reform Committee, U.S. House of Representative, *Corporate Information Security Working Group—Report of the Best Practices and Metrics Teams,* November 2004. Available at http://net.educause.edu/ir/library/pdf/CSD3661.pdf. Accessed July 19, 2012.

[10]   IEEE Standard 610.12-1990, *IEEE Standard Glossary of Software Engineering Terminology,* 1990. Available for purchase at http://standards.ieee.org/findstds/standard/610.12-1990.html. Accessed on July 19, 2012.

[11]   Jacquith, A., Security Metrics: Replacing Fear, Uncertainty, and Doubt, Upper Saddle River, NJ: Addison-Wesley, 2007.

[12]   See http://www.nyu.edu/pages/mathmol/textbook/weightvmass.html, last accessed on July 20, 2012.

[13]   See http://samate.nist.gov/index.php/Metrics_and_Measures.html, last accessed on July 20, 2012.

[14]   See http://www.riskglossary.com/link/risk_metric_and_risk_measure.htm   Accessed on July 20, 2012.

[15]   Munson, J. C., *Software Engineering Measurement,* Boca Raton, FL: Auerbach Publications, 2003.

[16]   Cruickshank, K. J., *A Validation Metrics Framework for Safety-Critical Software-Intensive Systems, Thesis,* Monterey, CA: Naval Postgraduate School, 2009. Available at http://www.dtic.mil/cgi-bin/GetTRDoc?AD=ADA496995. Accessed July 20, 2012.

[17]   See http://en.wikipedia.org/wiki/Software_metric.

[18]   McGraw, G., *Software Security—Building Security* In, Addison Wesley, 2006.

[19]   Braude, E. J., and M. E. Bernstein, *Software Engineering: Modern Approaches, Second Edition,* Hoboken, NJ: Wiley, 2011.

[20]   MISRA (The Motor Industry Software Reliability Association), *Development Guidelines for Vehicle Based Software,* Version 1.1, 1994. Accessed April 23, 2012 at http://www.ida.liu.se/~snt/teaching/SCRTS/MisraSoftwareGuidelines%5B1%5D.pdf

[21]   Anderson, R., *Security Engineering, Second Edition,* Indianapolis, IN: Wiley, 2008.

[22]   Herrmann, D. S., *Software Safety and Reliability,* Los Alamitos, CA: IEEE Computer Society, 1999.

[23]   Merkow, M. S., and L. Raghavan, *Secure and Resilient Software Development,* Boca Raton, FL: CRC Press, 2010.

[24]   Howard, M., and S. Lipner, *The Security Development Lifecycle,* Redmond, WA: Microsoft Press, 2006.

[25] Gutgarts, P. G., and A. Termin, "Security-Critical versus Safety-Critical Software," *Proceedings of the IEEE Homeland Security Technology Conference,* Waltham, MA, 2010, pp. 507–511.

[26] Axelrod, C. W., "The Need for Functional Security Testing," *STSC CrossTalk: The Journal of Defense Software Engineering,* Vol. 24, No. 2, March/April 2011, pp. 17–21. Available at http://www.crosstalkonline.org/storage/issue-archives/2011/201103/201103-0-Issue.pdf. Accessed July 20, 2012.

8

Software System Development Processes

> *For some decades, many development models, standards, laws, methods, assessments ... have been created to improve product quality ... The resulting quagmire becomes increasingly difficult to understand and almost inscrutable.*
> ——Dr. Winfried Russwurm, Siemens AG, *The Multi-Model and Multi-Appraisal Quagmire:An Approach to Organize by a Classification Scheme* [1]

Introduction

One aspect that is common to the majority of published approaches to software system development is the lack of attention to security and safety processes, which need to be incorporated into the basic set of process areas common to software development frameworks. This deficiency appears to be virtually universal and is made very apparent by those creating safety and security extensions to the basic processes.

Originally, there was no intention to spend much time on processes and process frameworks, with most of the focus to be on projects and project management. However, two events changed that view, resulting in more attention being given to processes. In fact, this entire chapter is devoted to processes. Projects and their management will be examined subsequently.

One motivator behind paying more attention to processes was the arrival of a prepublication draft of a report [2] that describes the +SECURE extensions to CMMI®-DEV [3]. These +SECURE extensions, together with the +SAFE

V1.2 extensions in [4], which are already available online, provide substantial guidance with respect to processes. These guidelines serve to encourage the development of safer and more secure software-intensive systems. While some documentation about safety and security extensions to the integrated CMM® already existed in Ibrahim [5] prior to the publication of the +SAFE and +SECURE extensions, focus on software safety and security will undoubtedly increase considerably with the advent of the +SECURE extensions [3] to complement the +SAFE document [4], which has been available since 2007.

At this point, it is worth noting the unusual chronology of the +SECURE draft document [2] and +SAFE V1.2 report [4]. In 2004, Ibrahim [5] referred to a +SAFE V1.0 document [6], which had been published in 2001. While it is true that the +SAFE V1.2 document [4] refers to the earlier +SAFE V1.0 document [6], it does so in the context that the latter "... was released to a limited audience for trial and evaluation." A point of interest is that neither the +SECURE draft [2] or the +SAFE V 1.2 report [4] mention this predecessor document [5].

Russwurm [1] has affirmed what many have suspected all along, namely that the whole area of process improvement is complex, convoluted, and difficult to navigate. While trying to cut through the maze of frameworks and models is too large a task for this book, this chapter should help explain the problem, even if it does not fully solve it. The hope is that there is value in examining the diversity of processes and process issues and the existence of inconsistencies across the various approaches and, by such means, improving the way in which safe and secure software systems are built and operated. However, excellent processes do not guarantee high-quality products and services, even though effective processes are more likely to generate products of better quality.

Processes and Their Optimization

It often seems, especially in the field of software development, that much time and effort is spent trying to improve processes. Similarly, the management of projects receives a great deal of attention. However, the amount of attention has not been enough to avoid many projects being late and over-budget. Furthermore, the resulting products and services often do not meet the requirements and expectations of stakeholders.

Much of the benefit from invoking world-class processes is lost if the resulting products and services are poor because the projects are badly managed. Furthermore, if even the best products do not go through an effective quality-assurance phase, the result is almost guaranteed to be poor quality products, services, and systems. That is not to say optimized processes are not important—they are. Mature, continuously-improving processes are generally considered to be necessary, though not sufficient, particularly for large multi-player projects. Nevertheless, using world-class processes does not necessarily ensure that end-products or services will satisfy stakeholders' requirements. Nor does it guarantee that products and services will be of sufficiently high quality, where quality includes attributes like security and safety. In fact, it seems that some researchers appear to be deluded into thinking that invoking a mature process will automatically produce quality products and services. However, it is entirely possible to build an inferior product using a superior process and good project management, particularly if safety and security requirements do not receive adequate attention or are completely omitted.

Notwithstanding the above disclaimers, there can be significant value generated from a complete set of well-organized processes. Unfortunately, as pointed out in [1], process models and frameworks are proliferating at breakneck speed, and it is a particularly difficult task to navigate through what Russwurm terms a quagmire. Furthermore, safety and security extensions to well-respected maturity models are relatively recent and, even within this subset of processes, we see the problems introducing these processes into entities that heretofore have not included specialty processes in their development life cycles.

In this chapter, we discuss some of the most commonly referenced process models, such as the CMMI® and various offshoots, particularly as they might relate to safety and security. We compare the processes for developing safety-critical and security-critical software systems. We also discuss how two seemingly-divergent philosophies might be brought closer together to address today's safety and security issues with systems of systems and cyber-physical systems, which must satisfy the requirements of many constituencies. This is not an easy task because the history of process models is complex and convoluted.

Processes in Relation to Projects and Products/Services

Even though organizations may achieve high levels of process maturity in some areas, not all projects necessarily fall under this umbrella. As stated in Chrissis [7] with regard to differentiating the CMMI®-DEV approach from others:

> ... most available improvement approaches focus on a specific part of the business and do not take a systemic approach ...

It is quite possible that some, or many, projects conducted in other parts of an organization may not reflect the quality and maturity of processes achieved by a particular area that, for example, obtains a high-level CMMI® certification. This is simply illustrated in Figure 8.1, where some projects are shown to fall within an established mature process whereas others do not. The process for obtaining certification requires significant time, effort, and cost, with the result

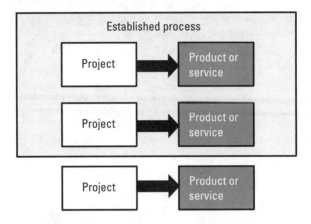

Figure 8.1 Projects within and outside mature process.

that only a few organizations could justify having all of their businesses and operational areas so certified. The problem resulting from this is that a customer
might presume that all products and services emanating from an organization
equally meet the maturity level asserted by a vendor, say, whereas not all products and services necessarily conform to this claim. This is a typical "buyer
beware" situation, in which it pays to perform appropriate due diligence.

A common theme running through this chapter is the differentiation between effectiveness and efficiency, where effectiveness is "doing the right things"
and efficiency is "doing things right or correctly." Thus, one can aim for *effective
processes* to provide greater assurance that projects falling under such processes
result in products and services that are wanted and expected by stakeholders.
A project that comprises a specific set of activities suggested by the process
must also be managed in an efficient manner. However, efficient projects do
not ensure that resulting products and services meet stakeholder requirements,
whether functional or nonfunctional. The only claim that can be made is that
the projects have been managed to make the best use of available resources.
Consequently, one should recognize that adhering to a well-defined process
may not be as useful as hoped for; this is so when projects that are governed by
the process, do not create products and services that meet all important stakeholder requirements (that is, the validation process) and fall within budget.

Effectiveness and efficiency are considered differently depending on
whether one is looking at these criteria from the process or project viewpoints.
For example, from a process-effectiveness perspective, it is important to have in
place appropriate specifications for the full range of process areas, whereas for
projects, effectiveness is achieved by obtaining complete and accurate specifications and validating that end-products and services satisfy those requirements.

Note that "process area" has a particular definition in the context of
CMMI®, as follows, per Chrissis [7]:

A cluster of related practices in an area that, when implemented collectively, satisfy a set of goals considered important from making improvement in that area ...

The key to efficient processes is in their appropriateness and acceptance within particular organizational cultures, whereas the efficiency of projects has to do with how they are planned and managed. As mentioned, assurance that the outputs of projects operating under a particular set of processes effectively meet stakeholder requirements is via *validation*, which mostly involves testing. The determination that, for each step of the development life cycle, the artifacts generated or components built in the prior phase do not contain defects is through *verification*, which generally consists of inspections, reviews and some testing. Whereas effectiveness and efficiency criteria affect both processes and projects, the emphasis is usually on the effectiveness of the former and the efficiency of the latter. Here, we will focus on the effectiveness of processes insofar as they are complete and appropriately directed. We will consider attaining project efficiency by means of good management.

In Table 8.1, we show how effectiveness and efficiency differ in terms of processes, projects and assurance.

Some Definitions

As usual, we will start out by examining the meanings of the words process, project, product, and organization, and not take them for granted or use one term where another one is more appropriate.

Sufficient distinction between process and project is often not made in the literature or in practice. This often leads to misinterpretations as to the meaning of certifications and reviews, such as CMMI® Level certification and SAS 70 Type I and Type II (now SSAE 16) audits [8].

It needs to be reemphasized that CMMI® and SSAE 16 examinations are usually restricted in scope to specific projects and operational areas respectively. However, as mentioned above, it is common for individuals who are not sufficiently aware of these limitations and restrictions to assume, albeit incorrectly, that the certifications apply across an entire organization. That is to say, a CMMI® Level 5 certification, which represents the highest and most mature

Table 8.1
Doing the Right Things and Doing Them Right

Criterion	Realization	Processes	Projects	Assurance
Doing the right thing	Effectiveness	Specifications	Requirements	Validation
Doing it right	Efficiency	Implementation	Management	Verification

level attainable, may have been obtained for a small subsidiary isolated from the main business, and yet one might be given to believe that it applies to the entire organization.

Similarly, SAS 70 (now SSAE 16) reviews are usually limited to a single operational unit. The reality is that it is not usually reasonable to jump from the particular to the general. Such leaps of faith inevitably lead to confusion, misinterpretations and inappropriate decisions and to the expectation that a level of excellence achieved by a subsidiary, say, applies across the board.

The following definitions come from the glossary in *CMMI® for Development* [3].

- A process is a "set of interrelated activities, which transform[s] inputs into outputs, to achieve a given purpose."

- A project is a "managed set of interrelated activities and resources, including people, [which] delivers one or more products or services to a customer or end user."

- A product is a "work product that is intended for delivery to a customer or end user."

It should be noted that the definition of a product according to the ISO/IEC 12207-2008 [9] is "the result of a *process*" [emphasis added]. This differs from the above definitions of process and project. In terms of the above definitions, a software product should be considered the result of a project, because a process is a description of what should be done in order to achieve a purpose, whereas products are created by a series of steps that makes up a project. This is another example where inconsistency across different sources results in confusion. The following definitions come from the glossary in *CMMI® for Development* [3]:

- A service is a "product that is intangible and non-storable."

- An organization is an "administrative structure in which people collectively manage one or more projects or work groups as a whole, share a senior manager, and operate under the same policies."

According to the above definitions, both processes and projects are sets of interrelated activities, except that a project is an instance of a process that is managed and which consumes specific resources used to create products or services.

The difference between a product and service has, according to the above definitions, to do with a service being intangible and nonstorable. Axelrod [10] describes services as being "perishable," in that services, which are not used, are no longer available for future use whether or not they were used in a prior period; this is the typical "use it or lose it" syndrome. It leads to the need to have excess service capacity to meet peak load conditions. This characteristic of perishability is also somewhat applicable to many products—typically those that flow, such as water, gas, oil, and electricity. These products do not disappear if not used—only the opportunity to use them during a particular time is often unable to be deferred, although some can be stored for future use. An example is a system by which excess hydro-electricity is used to pump water up to a lake, which is then used to produce electricity at some future time. If this is not done, then peak demand might exceed the capacity to generate. Many products, such as processed drinking water (as opposed to regular water supplies), are physically tangible, whereas others, such as electricity, are less tangible, especially if they are not visible in their native state. Nevertheless, all these products (where their delivery is in the form of services) can be measured and charged for based on the amount of product used or supplied within a given period of time.

However, even though services are not considered to be physical items *per se*, they often stem from the use of, or support for, physical items. It is really not so clear that services are intangible because they can result in measurable added value, as described above. In fact, services are the major component of modern developed countries' economies, where manufacturing and farm sectors have been relatively reduced as a percentage of total economic activity. Services usually accompany products, such as a meal (product) produced by a chef (project) and delivered by the wait staff (service), but they don't have to link to a product, as for example when one hires a babysitter. The value of a service can usually be inferred from the price someone is willing to pay for that service.

In the context of this book, we claim that a process consists of a series of steps, or set of interrelated activities, along with related policy, procedures, practices, and controls that are required to produce safe and secure software-intensive systems meeting appropriate and predetermined levels of quality. The achieved quality can be verified through formal inspection. In such a context, processes themselves don't actually produce anything; they define and govern projects that comprise activities that build a product or create a service. Processes can, and should, evolve and improve over time as experience is gained from projects, and flaws and inadequacies are identified and eliminated. Such improvement is the basis of the Capability Maturity Model Integration (CMMI®), originated in its current form and fostered by the Software Engineering Institute (SEI) at Carnegie Mellon University (CMU) in Pittsburgh, PA, with assistance from The MITRE Corporation, in 1986. It should be noted that SEI and The MITRE Corporation did not originate the concept of a step-by-step maturity model. More than a decade before the SEI/MITRE effort, Richard Nolan introduced the concept of stages in an organization's evolution [11].

According to Paulk [12], the purpose of the CMMI® framework is to "... help organizations improve their *software process*" [emphasis added]. In an overview, cited on the SEI website [13], the approach is defined as follows:

> CMMI® (Capability Maturity Model Integration) is a process improvement approach that provides organizations with the essential elements of effective processes, which will improve their performance.

Since 1986, there have been a number of major advances in maturity models with more recent releases extending the original concepts to the safety and security aspects of software system engineering processes.

Chronology of Maturity Models

The CMMI® process maturity framework was originally developed in the 1980s and formalized in 1993 in Paulk [12]. The CMMI® approach is specifically oriented towards software development and was used originally to assess "the ability of government contractors' *processes* to perform a contracted software project" [emphasis added] [14].

A number of additions to and variations from the original concentration on software development of CMMI® have occurred since the original version was created. CMU SEI has expanded the approach to cover many other organizational processes besides software development, such as those relating to acquisition, services, and people. A chronology of CMMs, specifically relating to safety and security, is shown in Table 8.2. It is derived from "Figure 1.2 The History of CMMs" in [3], but with some important additions, particularly regarding recent or pending publications; specifically, the +SECURE [2] and +SAFE [4] extensions to CMMI®-DEV [3].

Table 8.2
Chronology of Capability Maturity Models Related to Safety and Security

Document	Year	Derived from	Input to/Superseded by
+SAFE V1,0	2001	—	+SAFE v1.2
CMMI for Development V1.2	2006	CMMI for Development V1.1	CMMI for Acquisition V1.2 CMMI for Services V1.2 CMMI for Development V1.3 +SAFE V1.2
+SAFE V1.2	2007	+SAFE V1.0 CMMI for Development V.1.2	—
CMMI for Development V.1.3	2010	CMMI for Development V1.2	+SECURE V1.3
+SECURE V1.3 (Draft)	2012	CMMI for Development V1.3	—

The International Systems Security Association (ISSEA) used the CMM^{SM} approach to develop the Systems Security Engineering Capability Maturity Model (SSE-CMMI). The ISSEA published the SSE-CMM model in a guide in 2003, which five years later became ISO/IEC 21827:2008 [15,16].

REFERENCE ·

Security and Safety in Maturity Models

We will now examine several maturity models that have been created over the years.

FAA Model

In September 2004, the United States Federal Aviation Administration (FAA) published a report [5] on safety and security extensions for integrated capability maturity models. Almost half of the report is devoted to mapping the "safety and security application practices" against sources, and separately to safety and security sources. The application practices (APs) by goal are shown in Table 8.3.

It is interesting to note that neither safety nor security are specifically defined in Ibrahim [5], as is the case for many other similar reports—presumably

Table 8.3
Goals and Application Practices

Goals	AP Number	AP (Application Practices) Description
An infrastructure for safety and security is established and maintained	AP 01.01	Ensure safety and security competency
	AP 01.02	Establish qualified work environment
	AP 01.03	Ensure integrity of safety and security information
	AP 01.04.	Monitor operations and report incidents
	AP 01.05	Ensure business continuity
Safety and security risks are identified and managed	AP 01.06	Identify safety and security risks
	AP 01.07	Analyze and prioritize risks
	AP 01.08	Determine, implement and monitor risk mitigation plan
Safety and security requirements are satisfied	AP 01.09	Determine regulatory requirements, laws and standards
	AP 01.10	Develop and deploy safe and secure products and services
	AP 01.11	Objectively evaluate products
	AP 01.12	Establish safety and security assurance arguments
Activities and products are managed to achieve safety and security requirements and objectives	AP 01.13	Establish independent safety and security reporting
	AP 01.14	Establish a safety and security plan
	AP 01.15	Select and manage suppliers, products, and services
	AP 01.16	Monitor and control activities and products

one is supposed to gather the meaning of these terms from their use throughout the text. It is clear from [5] that safety and security are considered to be very separate areas, which is confirmed by such differences as, say, the sources used. Most of the safety-related sources are in the form of United States and United Kingdom military requirements and standards, whereas the security-related sources tend to be from global standards organizations. While such a separation of sources is to be expected, it nevertheless points to an obvious bias. This bias may partially explain why the military, up until quite recently, appeared to focus more on safety and less on security, and why fewer international standards (to date) have evolved for safety. This situation appears to be changing as the military pays greater attention to security as it applies to their information systems and networks, as well as to weapons systems and the like. Businesses are becoming more concerned about safety as applied to industrial control systems, which are increasingly being integrated into systems of systems and cyber-physical systems.

Ibrahim [5] states that the purpose of a so-called safety and security application area is to:

1. Establish and maintain a safety and security capability.
2. Define and manage requirements based on risks attributable to threats, hazards and vulnerabilities.
3. Assure that products and services are safe and secure throughout their life cycles.

As mentioned above, there are a number of comparisons made between the application area practices from [5] and various sources. Relevant safety sources, which are mainly military in origin, include the following:

- DEF STAN 0056: Safety Management Requirements for Defence Systems, U.K. Ministry of Defence, December 1996.

- MIL-STD-882C: System Safety Program Requirements, Military Standard, U.S. Department of Defense, January 1993.

- MIL-STD-882D: Standard Practice for System Safety, U.S. Department of Defense, February 2000.

- IEC 61508: Functional Safety of Electrical/Electronic/Programmable Electronic Systems, International Electrotechnical Commission, 1997.

Some relevant security sources, which tend to be international standards, are listed as follows:

- ISO/IEC 17799: Information Technology—Code of Practice for Information Security Management, International Organization for Standardization, 2000.
- ISO/IEC 15408: Common Criteria for Information Security Evaluation, Part 3: Security assurance requirements, Vol. 2.1, Common Criteria Project Sponsoring Organization, 1999.
- ISO/IEC 21827: System Security Engineering Capability Security Model (SSE-CMM), v3.0, SSE-CMM Project, 2003.
- NIST 800-30: Risk Management Guide for Information Technology Systems, Special Publication 800-30, National Institute of Standards and Technology, 2001.

The +SAFE V1.2 Extension

In 2007, the Defence Materiel Organisation of the Australian Department of Defence sponsored research by Carnegie Mellon University's Software Engineering Institute (CMU SEI). The result of this research was the +SAFE document, which is a safety extension to CMMI® for Development V1.2, (CMMI® -DEV), described in [7]. In 2010, Version 1.2 of the CMMI®-DEV was replaced with Version 1.3 [4].

The +SECURE V1.3 Extension

A group from Siemens AG developed a +SECURE extension [2] to the CMMI® for Development Version 1.3, a draft of which was made available to this author in March 2012. The final version of the report had not been released when this book was being written. However, much of the content of the forthcoming report is revealed in an article by Fichtinger [17].

We now examine CMMI® -DEV and its safety and security extensions in more detail in order to better understand how processes that incorporate safety and security process areas should be developed.

The CMMI® Approach

General CMMI®

According to CMMI® documentation, an organization evolves its processes through five stages or levels:

1. *Initial:* no structure or documentation to speak of.

2. *Repeatable:* the steps of the process are documented and may be repeated.

3. *Defined:* the process is considered to be a standard business process.

4. *Managed:* the process is managed against quantitative measures.

5. *Optimizing:* the process is continuously improved and optimized.

The range of processes included in the CMMI® portfolio is large and increasing as areas of engagement grow rapidly. There is a vast array of documents and training describing the CMMI® approach and how to obtain certification in various areas and at increasingly demanding levels. A good place to start to get a good understanding of the CMMI® approach is the CMU SEI website [18].

REFERENCE.

CMMI® for Development

CMMI®-DEV came out of the fundamental CMMI® process and more specifically addresses software development, even though the original intent of CMM was to improve software development efforts. Table 8.4 lists process areas within each CMMI®-DEV module or category, as provided in *CMMI® for Development, Version 1.3* [3].

ISO/IEC 12207-2008 [9] process groups are shown in Table 8.5. It should be noted that the processes cover both systems and software; there is another standard, ISO/IEC 15288, IEEE Std 15288-2008 [19], which only covers *system* life cycle processes, whereas both systems and software are covered in ISO/IEC 12207-2008.

We compare the entries in Tables 8.4 and 8.5 to one another in Table B.1 in Appendix B.

If you find the above confusing, you are not alone, as Russwurm so aptly stated in [1]. It is often difficult for someone with responsibility for the entire range of software development processes to determine which might be the best taxonomy to adopt. The more conservative approach is to take the superset of processes, which, in the above case, would be to take the process categories of ISO/IEC 12207-2008 and fill in any areas where CMMI®-DEV is more descriptive. The CMMI® approach is, as one would expect from the size of the documents alone (some 482 pages), much more comprehensive and provides guidance and support, including training. These types of support are not as readily available for the ISO/IEC standard.

Table 8.4

Specific Process Areas by Category per CMMI®

Process Area Category	Specific Process Areas
Process management	Organizational process focus Organizational process definition Organizational training Organizational process performance Organizational performance management
Project management	Project planning Project monitoring and control Supplier agreement management Integrated project management Risk management Quantitative project management
Engineering	Requirements development Requirements management Technical solution Product integration Product verification Validation
Support	Configuration management Process and product quality assurance Measurement and analysis Decision analysis and resolution Causal analysis and resolution

Incorporating Safety and Security Processes

A high-level comparison is given in Table 8.6 for those process areas with equivalent process (or application) areas as CMMI®-DEV in +SAFE V1.2 [4] +SECURE [2] and with the application areas in Ibrahim [5]. It is interesting to note that not every process area (PA) has an equivalent in the other documents. For example, there is no specific extension of the Process Management PA in the +SAFE document. The application practices (APs) in +SECURE [2] have equivalents, for the most part, in CMMI®-DEV and mapping that serve to illustrate the complexity of the quagmire of definitions and terms. Nevertheless, one can draw parallels, add, or join the dots in order to complete the picture. For example, there is clearly a need for organizational preparedness for safe development, which incorporates establishing an appropriate work environment and training individuals in particular skill sets, even though it may not be stated explicitly in the same terms across approaches.

We now look into +SAFE V1.2 [4] and +SECURE [2] in more detail.

+SAFE V1.2 Comparisons

Table 8.6 clearly shows that the safety extensions are in the CMMI®-DEV process areas of Project Management and Engineering, but that the process areas

Table 8.5
Life Cycle Process Groups per ISO/IEC 12207

Category	Subcategory	Individual Processes
System Context Processes	Agreement processes	Acquisition process Supply process
	Organizational project-enabling processes	Life cycle model management process Infrastructure management process Project portfolio management process Human resources management process Quality management process
	Project processes	Project planning process Project assessment and control process Decision management process Risk management process Configuration management process Information management process Measurement process
	Technical processes	Stakeholders requirements definition process System requirements analysis process System architectural design process Implementation process System integration process System qualification testing process Software installation process [distribution?] Software acceptance support process Software operation process Software maintenance process Software disposal process
Software Specific Processes	Software implementation processes	Software implementation process Software requirements analysis process Software architectural design process Software detailed design process Software construction process Software integration process Software qualification process
	Software support processes	Software documentation management process Software configuration management process Software quality assurance process Software verification process Software validation process Software review process Software audit process Software problem resolution process
	Software reuse processes	Domain engineering process Reuse asset management process Reuse program management process

of Process Management and Support do not appear to have equivalent safety-related processes. The +SAFE document [4] addresses these differences. The relationships among the process areas of safety management, safety engineering, and support are considered to be the same for +SAFE as for CMMI®-DEV. The +SAFE document does not appear to address the process management

Table 8.6
Mapping of CMMI®-DEV Process Areas to Safety and Security Extensions

CMMI®-DEV Process Areas	CMMI®-DEV Process-Area Subcategories	Safety and Security Extensions [5] Application Area	+SAFE [4] Process Areas	+SECURE [2] Process Areas
Process Management	Organizational process definition (OPD) Organizational training (OT)	Establish qualified work environment (AP 01.2) Ensure safety and security competency (AP 01.01)	—	Organizational preparedness for secure development (OPS)
Project Management	Project planning (PP) Supplier agreement management (SAM) Integrated project management (IPM)	Establish a safety and security plan (AP 01.14) Select and manage suppliers, products and services (AP 01.15) Establish independent safety and security reporting (AP 01.13)	Safety Management	Security management in projects (SMP)
Engineering	Requirements development (RD) Technical solution (TS) Verification (VER) Validation (VAL)	Determine regulatory requirements, laws and standards (AP 01.09) Develop and deploy safe and secure products and services (AP 01.10) Objectively evaluate products (AP 01.11)	Safety Engineering	Security requirements and technical solution (SRT) Security verification and validation (SVV)

process area, particularly the need for preparedness, including training of personnel. Presumably, this area reverts back to the parent document. However, it is something that should be made explicit for safety, especially as there is a lack of knowledge and expertise of safety engineering among developers of security-critical software and, as described in previous chapters, there is little interchange among the various silos of expertise. Training and knowledge transfer relating to safety are areas in need of particular explicit attention and the failure to call out these areas could well be a significant contributor to the current gap.

From +SAFE V1.2, it would appear that the CMMI®-DEV process for training is addressed in what is termed Specific Goal 1 (SG 1), although training is not mentioned in the description of SG 1.

Of particular interest is Table 4 in the +SAFE report which cross references +SAFE SGs with nine of the twenty-two CMMI®-DEV PAs. Table 8.7 summarizes the comparison at a higher level (i.e., not drilling down to individual SGs).

While this analysis may be comforting to some, there are questions as to how well the SGs match the CMMI®-DEV PAs and whether comparing in nine categories is adequate. Perhaps incorporating the +SAFE extensions directly into the parent document would have been a better approach, as it would clarify which process areas are addressed and which are not. Nevertheless, the +SAFE document [4] is a major step in the right direction despite its taking a safety perspective that is geared to safety engineers and may not be particularly helpful for others.

+SECURE V1.2 Comparisons

While the authors of the +SECURE document [2] assert that the +SECURE extension was "developed using the similar methods [to] the +SAFE extension

Table 8.7
Cross References of CMMI®-DEV vs. +SAFE Process Areas

CMMI®-DEV Process Areas	Safety Management Process Area	Safety Engineering Process Area
Causal analysis and resolution	Yes	—
Configuration management	—	Yes
Decision analysis and resolution	Yes	Yes
Measurement and analysis	Yes	—
Organizational training	Yes	—
Process and product quality assurance	—	Yes
Project monitoring and control	Yes	—
Project planning	Yes	—
Requirements development	Yes	Yes

...," the documents differ in many respects. We have already established that the +SAFE extension does not specifically address the process management process area, and that neither +SAFE nor +SECURE includes any support process areas. However, there are many other differences that show up as one digs down into the specific goal and practice levels, as illustrated in Table 8.8.

Table 8.8 is very revealing. First of all, it shows that the +SECURE specific goals and practices cover many more areas than do the specific goals of the +SAFE extension and the +SECURE specific goals appear to be much more oriented to the software development process than are the safety goals and practices. The safety approach appears to be less formal than the security approach. For example, in the design category, the +SAFE requirement is to apply safety principles, whereas +SECURE calls for using security standards.

Summary and Conclusions

In many ways, this venture into the world of processes was highly informative for the author, and, it is hoped, for the reader, too. The overall impression matches Russwurm's representation of software development process models as a quagmire [1], which was unfortunately shown to be accurate. One who is interested in implementing one of the many process frameworks is sure to be confronted with a myriad of approaches. These approaches are difficult to compare and each requires substantial effort to understand and implement.

Attempts to extend process models from the general software development framework to safety and security are highly variable in their rigor and usefulness. There also appears to be a tendency to reinvent the wheel to the extent that references to prior work and to parent documents are often incomplete.

Nevertheless, for all their problems, process improvement methods are much needed, as are those that relate specifically to safety and security. It is crucial to include safety and security in process models because, if they are not included, there is little hope for advancing the art of secure and safe software systems engineering. That being said, the process models are not yet good enough to assure IT professionals that the complex systems that they build and operate will meet the security and safety requirements of diverse stakeholders.

Whereas world-class processes do not guarantee top-quality products and services, the lack of adequate processes practically ensures that many critical considerations relating to safety and security will not be included in the overall processes and projects that underlie the creation of all software-intensive products and software-based services.

Table 8.8
Comparison of Specific Goals and Specific Practices for +SAFE and +SECURE

+SAFE Specific Goals	+SAFE Specific Practices	+SECURE Specific Goals	+SECURE Specific Practices
—	—	Establish capabilities to develop secure products and to react to product security incidents	Obtain management commitment, sponsorship and involvement Establish standard processes and other process assets for secure development Establish awareness and capability for product security Establish secure work environment standards Establish vulnerability handling
—	—	Prepare and manage project activities for security	Establish integrated project plan for security projects Plan and deliver security training Supplier and third party component selection for secure products Identify underlying causes of threats
Identify hazards, accidents, and sources of hazards	Identify possible accidents and sources of hazard Identify possible hazards	—	—
Analyze hazards and perform risk assessments	Analyze hazards and assess risk	Manage product security risks	Establish product security risk management plan Perform product security risk analysis Plan risk mitigation for product security
Define and maintain safety requirements	Determine safety requirements Determine a safety target for each safety requirement Allocate safety requirements to components	Develop customer security requirements and secure architecture and design	Develop customer security requirements Design the product according to secure architecture and security design principles Select appropriate technologies Establish standards for secure product configuration
Design for safety	Apply safety principles Collect safety assurance evidence Perform safety impact analysis on changes	Implement the secure design	Use security standards for implementation Add security to the product support documentation
—	—	Perform security verification	Prepare for security verification Perform security verification
Support safety acceptance	Establish a hazard log Develop a safety case argument Validate product safety for the intended operating role Perform independent evaluations	Perform security validation	Prepare for security validation Perform security validation

Endnotes

[1]   Russwurm, W., "The Multi-Model and Multi-Appraisal Quagmire—An Approach to Organize by a Classification Scheme," *SEPG 2010*, Atlanta, GA, 2010, available at http://www.sei.cmu.edu/sepg/na/2010/. Accessed July 20, 2012. (Note: The presentation can be reached via clicking on the link "Thursday's Sessions," and then clicking on "1541_Russwurm.")

[2]   Siemens, A. G, *+SECURE, V1.3: A Security Extension to CMMI-DEV, Vol. 3*, Software Engineering Institute, March 2012 (currently in draft—limited distribution).

[3]   Software Engineering Institute, *CMMI® for Development, Version 1.3, Technical Report CMU/SEI-2010-TR-033 ESC-TR-2010-033, Improving Processes for Developing Better Products and Services,* November 2010. Available at http://www.sei.cmu.edu/reports/10tr033.pdf. Accessed July 20, 2012.

[4]   Defence Materiel Organisation, Australian Department of Defence, *+SAFE, V1.2: A Safety Extension to CMMI® -DEV, V1.2, Technical Note CMU/SEI-2007-TN-006,* Software Engineering Institute, March 2007. Available at http://www.sei.cmu.edu/reports/07tn006.pdf. Accessed July 20, 2012.

[5]   Ibrahim, L., et al., *Safety and Security Extensions for Integrated Capability Maturity Models,* United States Federal Aviation Administration, 2004. Available at http://www.faa.gov/about/office_org/headquarters_offices/aio/library/media/SafetyandSecurityExt-FINAL-web.pdf. Accessed July 20, 2012.

[6]   SVRC (Software Verification Research Centre) Services, *+SAFE, A Safety Extension to CMMI*, UniQuest Pty Lmt, Queensland, Australia, 2001.

[7]   Chrissis, M. B., M. Konrad, and S. Shrum, *CMMI® for Development, Version 1.2, CMMI® Second Edition: Guidelines for Process Integration and Product Improvement,* Upper Saddle River, NJ: Addison Wesley, 2007.

[8]   See http://www.ssae16.org/white-papers/ssae-16-vs-sas-70--what-you-need-to-know-and-why.html, last accessed on July 20, 2012.

[9]   Software & Systems Engineering Standards Committee of the IEEE Computer Society, *International Standard ISO/IEC 12207 IEEE Std 12207-2008: Systems and Software Engineering—Software Life Cycle Processes,* ISO/IEC/IEEE, 2008.

[10]   Axelrod, C. W., *Computer Effectiveness: Bridging the Management/Technology Gap*, Washington, DC: Information Resources Press, 1979.

[11]   See Nolan, R. L., "Managing the computer resource: a stage hypothesis," *Communications of the ACM,* Vol. 16, Issue 7, July 1973. Available for purchase at http://dl.acm.org/citation.cfm?doid=362280.362284, last accessed on July 20, 2012. [12]   Paulk,   M. C., B. Curtis, M. B. Chrissis, and C. V. Weber, *Capability Maturity Model SM for Software, Version 1.1, Technical Report CMU/SEI-93-TR-024 ESC-TR-93-177,* 1993. Available at http://www.sei.cmu.edu/reports/93tr024.pdf. Accessed July 20, 2012.

[12]   Paulk, M. C., B. Curtis, M. B. Chrissis, and C. V. Weber, *Capability Maturity Model SM for Software, Version 1.1, Technical Report CMU/SEI-93-TR-024 ESC-TR-93-177,* 1993. Available at http://www.sei.cmu.edu/reports/93tr024.pdf, last accessed July 20, 2012.

[13]   See http://www.sei.cmu.edu/cmmi/, last accessed on July 20, 2012.

9

Secure SSDLC Projects in Greater Detail

A comprehensive, systematic approach to implementing security from the very start of applications development is essential.
—Patrick McBride and Edward P. Moser, [1]

In an era riddled with asymmetric cyber attacks, claims about system reliability, integrity and safety *must also include provisions for built-in security of the enabling software* [emphasis added].
—The Data and Analysis Center for Software (DACS), Department of Defense [2]

Dependability specifications are analogous to safety and security requirements ...
—Praxis High Integrity Systems, [3]

Introduction

As we have already discussed, there are numerous acronyms for security-critical and safety-critical software-intensive system-development life cycles. Many of these acronyms, though seemingly different, apply to very much the same types of systems and processes, whereas others are clearly or subtly different, often ambiguous. All such attempts to define the development process and life cycle components are intended to bring structure, formality, completeness, and repeatability to processes and procedures, which are often ad hoc in practice. Many processes address only the creation of standalone software products; others take into account broader aspects of software-intensive systems by including platforms, infrastructures, and possibly hardware. Yet other approaches specialize in one attribute of software such as security, safety, reliability, integrity, availability, performance, resiliency, and the like, but pay little attention to the other characteristics of software systems. Thus books about software security seldom mention safety and those covering the topic of software safety pay little or no attention to security.

The purpose of this chapter is to examine the phases and detailed tasks and subtasks that are most crucial to developing, operating and decommissioning secure software systems. In Chapter 10, we shall look into the same processes as they apply to safety-critical software systems. We shall then look at the particular issues that relate to creating software systems that are both security-critical and safety critical.

Different Terms, Same or Different Meanings

The simple acronym SDLC is itself ambiguous. In some instances, SDLC refers to the *system development life cycle*, or *system design life cycle*, whereas in other cases it has the meaning *software development life cycle*. Going back some years, the IBM Corporation came up with the abbreviation SDLC for *Synchronous Data Link Control*, referring to a network communications protocol. To further complicate matters, Microsoft Corporation has adopted the acronym SDL to stand for the "security development life cycle" [4], when the authors really meant "secure software development life cycle."

In this book, we will use the acronym SSDLC to mean software system development lifecycle. It should be noted that this abbreviation is neither original nor unique; it has been used to mean secure software development life cycle in a number of cases [5]. Others used SSDLC to mean "security system development life cycle," which has the same problem of confusing secure systems with security systems, as discussed previously [6]

We shall depict particular flavors of the SSDLC using the terms *secure SSDLC, safe SSDLC, resilient SSDLC,* and the like. This is because the term software system emphasizes the fact that software is not developed in isolation, nor can it run without other software, firmware, and hardware components. Software is always part of a broader system that includes platforms (i.e., operating systems, system utilities) and infrastructures (i.e., equipment and networks), and must generally be designed and developed with specific platforms and infrastructures in mind.

In order to accommodate security and safety, we use terms such as secure software system development lifecycle (secure SSDLC) and safe software system development lifecycle (safe SSDLC). When we consider building systems that have both explicit security and safety requirements, we use the cumbersome term safe and secure software system development life cycle (S⁴DLC), favoring precision over brevity.

At this point, we will again differentiate among secure software systems, security software systems, and secure security software systems. With respect to secure software systems, security is a quality *attribute* of the software systems, whereas for security software systems, security is the predominant *function* of the software systems. Odd as it may seem, the term secure security software system is not redundant because it refers to software systems, the main function of which is to provide security services. The attack against RSA's SecureID technology, detailed below, is an example of the security of a security system being compromised. Therefore, these same security software systems need be built so that the security software systems themselves exhibit strong security attributes and are resistant to attack. The reported attacks against RSA and Symantec demonstrate that these security product vendors are perhaps more vulnerable than generally understood [7].

The terms and meanings are similar for safety-critical systems and combined security-critical and safety-critical systems. As with secure software systems, safe software systems are those for which safety is an *attribute*. For safety software systems, safety is the main *function* of the software systems. Therefore, safe safety software systems provide safety services and themselves meet specified safety standards. Secure safety systems are those for which the main function is safety, but which are also secure because they are protected from adverse attacks or events. Safe security systems are those that provide physical or logical security functions and do not threaten harm to people, other creatures, or the environment. An example of the latter would be a building entry system that fails in an open state so that, in the event of a fire, say, persons within the facility can escape and firefighters, emergency service personnel, and other responders can enter the facility freely.

Even more confusing are safe and secure software systems that reflect safety and security requirements, but have functions other than the provision of safety and security services.

When we talk about safe and secure safety and security software systems, we mean software systems that function to preserve both safety and security, and also meet specified safety and security requirements. Thus, one might have (as with systems of systems and cyber-physical systems, such as the smart grid [8]) safety-critical systems that satisfy safety standards and have been secured against outside attacks and internal mishandling. Conversely, such systems may be those that provide security services but have also been specified as not doing damage to living beings or the environment were they to fail or malfunction. The above is summarized in Table 9.1.

Table 9.1 illustrates how the attribute of security (which previously was not considered a high priority for safety-critical systems) is now considered to be important, particularly for systems that have both security and safety requirements. Security-critical systems clearly must exhibit high degrees of intrinsic

Table 9.1
Safety and Security Attributes vs. Safety and Security Functions

Attributes	Functions		
	Safety	**Security**	**Combined Safety & Security**
Security	Currently high Traditionally low	Very high	Very high
Safety	High	Currently medium Traditionally low	Very high

security due in part to the negative reputational fallout when compromised, as in the case of the security-firm RSA, which was the victim of a successful attack that brought into serious question the security of their two-factor authentication technology [9].

RSA sells one of the most popular two-factor authentication technologies in the form of their SecurID tag, which is usually in the form of a key fob or as credit-card size unit with a small screen on it. The screen displays a numeric code which changes every minute and is synchronized with a number generator on the system to which access is to be granted. The use of the SecurID card, in conjunction with some other code, such as a password or personal identification number (PIN) provides two-factor authentication, which is considered to be highly secure. As a result, SecurID technology is used to protect some of the most security-critical software systems, such as those operated by defense contractors. When RSA was compromised by a cyber attack, the underlying SecurID technology was stolen. The technology was then used to attempt to infiltrate the systems of such entities as large defense contractor Lockheed Martin. From the perspective of cybersecurity professionals, this compromise was particularly disconcerting because it showed weaknesses in what was considered to have been one of the most effective ways to prevent unauthorized access to critical systems.

With respect to safety, a system controlling something like entry to a high-security facility needs to have very good safety features. If the safety software were to fail, the reputation of the company supplying the control software and its ability to perform would be brought into question. When systems of systems and cyber-physical systems have both safety and security requirements, the controlling software must have particularly stringent safety and security attributes.

Creating and Using Software Systems

There are many published approaches to building software systems. Most of them follow the traditional basic systems engineering approach to projects. Some expand beyond the basic scope and add specific security and safety com-

ADD DISPOSAL ≥ DECOMMISSIONING TO PROJECT

ponents. Others are less complete than the basic systems approach. For example, many software system lifecycles omit consideration of disposal or decommissioning of obsolete systems and planning for their replacements with new systems. Furthermore, typical development life cycles frequently do not follow the commonly-held of birth-life-death-rebirth cycles, restricting consideration to a single pass. The inevitable end-of-life of software systems is often not planned, particularly for security-related lifecycle processes in the private sector. To their credit, many government approaches do consider the decommissioning and disposal of secure software systems, although the decommissioning and disposal phases appear more often in safety-related control systems than they do for information systems. It is emphasized here that the activities that take place when a software-intensive software system is discontinued must be carefully planned if one is not to be exposed to a litany of risks, including loss of valuable intellectual property. Many researchers and practitioners appear to assume that once the useful life of a software system is at an end, then the software has no value. Often this is not the case because older versions of software products, while not necessarily supported by the vendor, can usually still operate usefully, if so desired and may well contain valuable intellectual property.

In some cases, even when a software system is decommissioned, the data handled by the system may have to be retained over a longer term so as to meet the requirements of the taxing authorities or regulatory agencies. Sometimes, data records must be retained for a given number of years; in other cases, they may have to be retained forever. This raises a key issue regarding the ability to retrieve electronic records that are compatible with the decommissioned software systems. If replacement systems are not backward-compatible, then some means to reconstitute the replaced software and hardware systems may ASSIGNMENT be required in order to retrieve required data. This can become a particularly onerous requirement, especially if the systems and those familiar with the applications and those who are needed to support the systems operationally are long since gone.

In addition, there are seldom preliminary plans for the creation of a replacement system put in place at the time the original system is created. Individuals developing software systems usually take no responsibility for consideration of replacement systems because it only takes more time and resources to include features that will facilitate replacing the original system; expending those resources can only be detrimental to the timely completion of the project at hand, unless consideration of replacement is a specific requirement.

This omission of replacement considerations can lead to panic and confusion when, for example, a vendor announces that it will no longer support a specific software product beyond a given date, or if the business conditions change drastically, calling for a different system, be it needing greater scalability, increased functionality, and the like, which may well be beyond the capabilities of the current software system.

Between 2004 and 2005, the Microsoft Corporation announced that it would no longer support Windows NT 4.0 beyond a specific date [10]. This created a huge uproar among customers who were not able to upgrade to a new system within the remaining time. Microsoft then agreed to provide security patches; however, they wanted a substantial additional maintenance fee in exchange for these patches. This further angered customers who felt that Microsoft was forcing them into changes and support terms and conditions that they felt were onerous.

Phases and Steps of the SSDLC

We will now examine each phase, and the steps within each phase, of a typical SSDLC, with particular initial focus on the security attributes of those systems, and consideration of the safety attributes in Chapter 10.

In Figure 9.1, we show three aspects of the life cycle inside the ovals. The center stream, labeled "Manufacture," is the typical sequence of development activities. In parallel with the development of the system is an "Oversight," or managerial cycle, which involves the initial risk analysis and continuing audit reviews and testing by auditors and examiners. The bottom third of the diagram shows the "Assurance" cycle, which comprises developing test cases and test plans and performing the various forms of testing. The shaded boxes in the diagram are those functions that are often not given the amount of attention commensurate with their importance. Two areas stand out as much-needed enhancements to the typical SSDLC. One is the specification of nonfunctional requirements and their testing to ensure that they are correctly incorporated into the system. The other is the decommissioning and disposal phases that need to be planned for, tested, and performed, but are often neglected in the private sector in particular.

A somewhat different view of the same set of concepts is shown in Figure 9.2, which derives largely from McGraw [11]. Here, the security aspects are related specifically to phases of the SSDLC.

It is interesting to review how the SSDLC has evolved. We now examine what particular steps were considered to be part of the system development life cycle around 1980. Historically, researchers did not usually make explicit distinctions between systems processes in general and software systems, processes, and projects in particular. In Biggs [12], there is a very detailed description of what we might term *software-intensive* systems *project* management rather than the process management of systems, as implied by the title of the book, which is *Managing the Systems Development Process.*

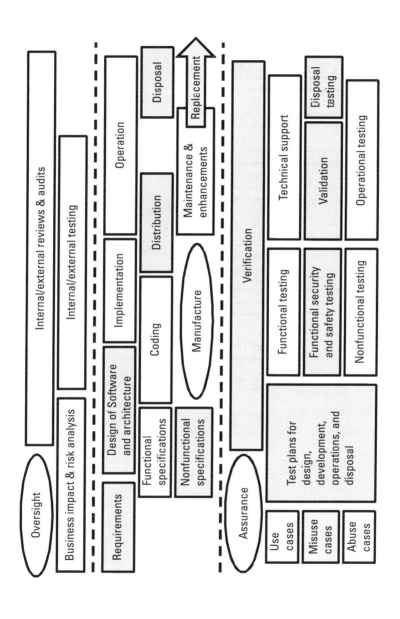

Figure 9.1   The SSDLC from the oversight, manufacture and assurance perspectives.

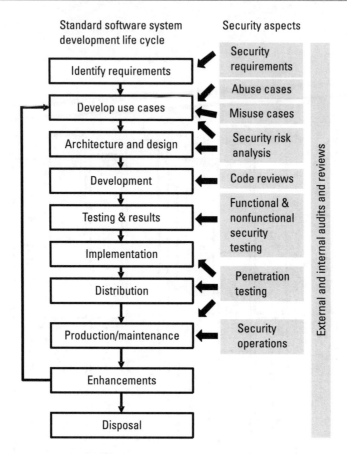

Figure 9.2 Security aspects related to the SSDLC.

Table 9.2 provides an overview of the various phases and steps of the development life cycle as described in [12]. It is noteworthy that there is no mention of security or safety and that there is no decommissioning phase. The approach does, however, require reviews by senior management and approval of the prior phase before the subsequent phase is initiated. Also, in Phase III: Systems Development, we see the term specifications relating to the technical aspects of the system and with respect to the application (Steps 1 and 3, respectively). While the term was used commonly in the 1980s, it appears to have somewhat gone out of favor in texts relating to the secure development of software systems. The content relative to specifications is generally subsumed under the requirements analysis or the architecture and design phases.

As an example, in the U.S. Federal Aviation Administration's report on the SDLC [13], under the heading "Software requirements analysis," we read the following:

Table 9.2
Traditional Systems Development Process per Biggs [6]

	Phases	**Steps**
Phase I	Systems planning	Step 1: Initial investigation
		Step 2: Feasibility study
	Top management review and decision	
Phase II	Systems requirements	Step 1: Operations and systems analysis
		Step 2: User requirements
		Step 3: Technical support approach
		Step 4: Conceptual design and package review
		Step 5: Alternatives evaluation and development planning
	Top management review and decision	
Phase III	Systems development	Step 1: Systems technical specifications
		Step 2: Technical support development
		Step 3: Application specifications
		Step 4: Application programming and testing
		Step 5: User procedures and controls
		Step 6: User training
		Step 7: Implementation planning
		Step 8: Conversion planning
		Step 9: Systems test
	Top management review and decision	
Phase IV	Systems implementation	Step 1: Conversion and phased implementation
		Step 2: Refinement and tuning
		Step 3: Post-implementation review
	Systems maintenance	Step 1: Ongoing maintenance

- *Safety specifications*: includes those relating to methods of operation and maintenance, environmental influences, and personal injury

- *Security specifications*: includes those relating to the compromise of sensitive information.

These descriptions differentiate between safety and security specifications in much the same way as we have done earlier.

Within each step, Biggs [12] provides an overview and lists of standard activities and key considerations. For example, in Phase III, Step 4, Application Programming and Testing, we have the following standard activities:

- Verify the availability of all necessary resources;

- Assign applications to programmer(s) and monitor progress;
- Develop program;
  - Code;
  - Compile;
  - Unit test (module or subroutine);
  - Integrate and test;
  - Document;
- Review and approve test results and documentation;

Also in Phase III, Step 4, Application Programming and Testing, we have the following key considerations:

- Know the software;
- Control of creativity;
- The 90% complete program;
- "Impossible" conditions;
- Resource utilization.

These activities and considerations are what one would expect for this step of a typical software system development project. However, an item of particular interest to considerations of security and safety (although not mentioned specifically), is the one labeled as "impossible conditions." As described by Biggs, this condition "happens because a situation is encountered that users, systems analysts, and programmers thought 'impossible'" [12]. According to Biggs, such conditions are not preventable, but they can be avoided or minimized to the extent that they can be predicted or anticipated or, if they do occur, then somewhat more timely corrective action can be taken [12].

Biggs proposes the use of structured walkthroughs and reviews to attempt to anticipate these conditions and from there develop mitigating procedures [12]. This is somewhat analogous to the concept of *functional security testing*, as described in Axelrod [14], or *functional safety testing*, which is the analogous method with respect to safety-critical software systems. In retrospect, it is strange that safety and, in particular, security considerations were not even mentioned in earlier work. In a sense, this lack of attention to such matters in prior decades might well have contributed to the failure to anticipate the need to build safety, and particularly security, into software systems.

The initial point of entry for the development of safe and secure systems is indeed the requirements phase. In fact, many writers believe that deficiencies in requirements are the main contributors to omissions and errors in software systems, many software systems development efforts fail to fully satisfy stakeholders' needs and often fall far short of stakeholders' expectations because of inadequate requirements.

Undoubtedly, a key critical success factor is obtaining a complete, comprehensive, unambiguous, and doable set of requirements, in which safety and security issues have been addressed early in the development lifecycle and continue to be addressed throughout the lifecycle, up to and including the decommissioning and disposal of the system. A more detailed description of the essential attributes of requirements is as follows:

- *Complete and comprehensive requirements:* these include material from all stakeholders (including requirements for security, safety, business continuity, performance, etc). One way to improve the chances of getting all relevant requirements is to think of all the various groups that may have a vested interest in the system development effort succeeding—and also of the groups that may be looking for the project to fail. This is often embodied in the "nobody asked me" syndrome, wherein information is not properly conveyed because it was not asked for. There is often a need to motivate stakeholders to respond and to be forthright in their responses. This can often be achieved by encouraging stakeholders to participate through various incentives, such as positive periodic personal reviews, special mention in company newsletters, and other forms of recognition. The effectiveness of the methods chosen will depend upon the culture of the organization. Suggested involvements for each phase of the life cycle are shown in Appendix C for security-critical software systems. ᴀꜱꜱɪɢɴᴍᴇɴᴛ

- *Unambiguous requirements:* it is important for stakeholders to under- 1 stand the terminology being used and to be able to translate their personal needs into highly specific requirements. Those requirements, both functional and nonfunctional, are then converted into specifications, which are used as the basis for system design and architecture, coding, testing, and the like. At each stage, there needs to be confirmation that the program code accurately implements requirements, whether or not the requirements truly represent the needs of the user. This is the verification process. In addition, the originator of the requirements (usually the customer or user) should make sure that what has been produced byt the project team is the right product (that is, it is what the user expected to see). This is the validation process.

- *Doable requirements:* in all cases, requirements need to be reasonable and achievable with existing technology (or allowance has to be made if technology is new or very different from the norm). Sometimes it might not be known until well into the development effort whether or not required capabilities are achievable within reasonable time and money constraints. The ability to determine feasibility without major expenditures and effort is the value of techniques such as prototyping, agile development, and the like. Many stakeholders, particularly among end users, often do not understand the terms used by technologists and their implications unless they actually see a model or prototype of the system. Not wanting to appear ignorant, some stakeholders may allow the process to continue, whether or not there are definite warning signs that the project might be going off the rails.

Given the above, we see that today safety and security requirements tend to be different animals. Safety requirements are usually based on specific standards, such as DO-178B, which will be discussed in more detail in Chapter 10. Security requirements, on the other hand, frequently do not have an equivalent set of standards and certifications to turn to. They are often based on de facto standards, such as the Payment Card Industry Data Security Standards (PCI DSS), which specifically addresses credit cards, debit cards, and other forms of payment systems, such as smart phones, which are increasingly being used in place of plastic cards.

There are also high-level management standards for security, such as ISO/IEC 27001/27002, which provide guidance for implementing security measures.

Calder [15] provides a good guide to certification under the ISO/IEC 27001/27002 standards. The areas covered by the standards are extensive, but adherence to these standards is usually voluntary, so that their value is diminished. However, it is helpful to be aware of the range of subject areas that are covered, which are as follows:

- *Organizing information security;*
- *Information security policy and scope; risk assessment;*
- External parties;
- Asset management;
- Human resources security;
- Physical and environmental security;
- *Equipment security;*
- Communications and operations management;
- *Controls against malicious software;*
- Network security management and media handling;
- Exchanges of information;

- E-mail and internet use;
- *Access control* (networks, *operating systems, applications*);
- *Systems acquisition, development and maintenance;*
- Cryptographic controls;
- *Security in development and support services;*
- *Monitoring and information security incident management;*
- *Business continuity management;*
- Compliance.

The above areas in italicized print have particular relevance to software engineering for security-critical applications. The other areas are important for a comprehensive information security program.

There was also an effort to develop a set of Generally-Accepted Information Security Principles (GAISP). An initial effort, begun in 1992, was called Generally-Accepted System Security Principles (GASSP). A description of the GASSP was issued for public comment purposes in 1999 [16]. The GASSP project was originally sponsored by the International Information Security Foundation (I²SF). The Information Systems Security Association (ISSA) took over the effort with a project involving more than a hundred volunteers and renamed it the Generally-Accepted Information Security Principles (GAISP) project [17]. Unfortunately, the GAISP effort collapsed under its own weight, with the net result that there is still no set of universal security principles and standards. In contrast, the American Institute of Certified Public Accountants (AICPA) and the Canadian Institute of Chartered Accountants (CICA), with assistance from the Information Security and Control Association (ISACA) and the Institute of Internal Auditors (IIA)), developed the Generally-Accepted Privacy Principles (GAPP) with a relatively small team of volunteers and published the results in two editions, the Business Edition [18] and the Practitioner Edition [19].

There are very specific software security recommendations, such as the Open Web Application Security Project (OWASP) top ten vulnerabilities and the SANS top 25 errors made by programmers described in Appendix A, as well as work done by The MITRE Group on a Common Vulnerability Scoring System (CVSS) [20] and a Common Weaknesses Evaluation (CWE) [21]. However, the guidelines for security do not even approach the rigorous certification requirements of safety-critical systems.

The net result of all this complex tapestry of security standards is a quandary which many designers, developers, and testers of security-critical software systems face, regarding what they need to do to ensure that the security attributes of a software system meets generally accepted standards. Because no such security standards exist, the approach recommended here is to examine each security-related task and subtask within the development life cycle and determine what makes sense to do at each stage and for what purpose.

The tables in Appendix C show security-related task areas, subtasks, and purposes of those subtasks for the following software systems development life cycle phases:

- Requirements analysis;
- Architecture and design;
- Development (building/configuration/integration);
- Testing;
- Deployment;
- Operations/maintenance.

It should be noted that specifications and decommissioning phases are not included explicitly in the above list; they should be added either as specific task areas or incorporated into existing tasks. It seems that including explicit specification tasks or subtasks has fallen out of favor in references covering the secure SSDLC. A *specification* is a detailed description of the design and materials used to make something. In Table 9.2, above, there are two types of specification—technical and application. On the other hand a *requirement* is a "singular documented physical and functional need that a particular product or service must be or perform"[22]. That is to say, a requirement is an expression as to *what* is wanted, whereas a specification describes *how* the product should be manufactured or the service developed. It is reasonable, therefore, to include specifications as an extension to the design and architecture phases, although Biggs [12] puts application and technical specifications in the development phase.

On the other hand, the decommissioning phase can be considered separately even though it might be considered to be an extreme example of a system modification. With respect to omitting the decommissioning phase, it is unfortunate that this phase is so often neglected or omitted completely, particularly as it is arguably a period of very high risk with respect to changing procedures and operations with a relatively high risk of unauthorized access to data and the misuse of sensitive information.

Detailed security requirements and specifications provide much more useful information for the builders of security-critical systems than do the high-level requirements. However, the high-level requirements are more easily understood by nontechnical participants (i.e., customers, clients, etc). Drilling down deeper, we see that detailed tasks and subtasks provide very specific guidelines for secure coding, for example. These latter guidelines are for shirtsleeve developers, testers, and implementers, but also need to be understood by project managers. The need to utilize secure methods has to also appeal to business-unit managers as a means of improving their products and services. Microsoft is a good example of this, as evidenced by their development life cycle described by Howard [4].

In Appendix C, we examine each identified phase of the development life cycle, and show a full range of security-related task areas, specific subtasks, and the reasons for performing each subtask. Much of the information contained in these tables comes from MacBride [1], which is not readily available [23]. However, the decommissioning/disposal phase has been added as have a number of the individual task areas and subtasks. Also, the format of the tables and how the information is distributed throughout the tables is new.

Table 9.3 shows the overall involvement of the various participating groups and teams in the security aspects of the lifecycle. The table is taken from Table C.17 in Appendix C. As expected, the major overall participants are the security and audit team members, but the application development team and the infrastructure team follow close behind. For the effort to be successful, participants must be fully engaged and perform the tasks assigned to them diligently.

What conclusions might one draw from Table 9.3? It would appear that the security team and the internal and external auditors should be heavily engaged throughout the lifecycle and that most of the other players hover around the moderate value, which is what one might expect. However, in a typical organization, we do not see this level of involvement. Most areas would be assessed at least one level lower in terms of their involvement. This is even true of security teams and auditors, whom you would expect to be very much in tune with the Secure SSDLC. The level of involvement is to some extent represented in the Building Security in Maturity Model (BSIMM) described at the BSIMM website [24].

Summary and Conclusions

In order to ensure that appropriate security attributes are included in software systems, especially those that are considered to be security-critical, it is necessary to include tasks and subtasks within the development lifecycle that are specifically geared to meet security requirements. Unfortunately, there is a dearth of information security principles, and detailed standards available to software safety engineers. Some guidance is available from various government and industry sources, and there are international standards that provide high-level recommendations for those wanting to be certified under those standards. How-

Table 9.3
Summary of Levels of Overall Involvement by Phase

Phase	Overall Level of Team/Group Involvement							
	Security Team	Application Development Team	Infrastructure Team	Project Managers	Business Unit Managers	Operations Managers	Internal and External Audit Teams	Executive Management
Requirements	High	Low to moderate	Low to moderate	Low to moderate	Moderate to high	Moderate to high	High	Moderate to high
Architecture and Design	High	High	High	Moderate to high	Moderate to high	Moderate	High	Low to moderate
Development	High	High	High	High	Moderate	Moderate	High	Low to moderate
Testing	High	High	High	High	Moderate	Low	High	Low
Deployment	High	Moderate to high	Moderate to High	High	Moderate	Moderate	High	Moderate
Operations and Maintenance	High	Moderate	Moderate	Moderate	Moderate to high	Moderate	High	Moderate
Decommissioning	High	Low	Low	Low	Low	High	High	Low
Overall Life Cycle	High	Moderate to high	Moderate to high	Moderate to high	Moderate to high	Moderate	High	Low to moderate

ever, many entities are not required to follow any particular standards, resulting in an abundance of weaknesses. Even if there were a set of generally-accepted standards, there would still be a need to engage internal and external subject-matter experts in the broad range of information security areas. (*)

One cannot hope to achieve the required level of security if those who are responsible do not engage in the process and ensure that all necessary steps are taken. The involvement of auditors is also required to ensure that these steps have been completed at a satisfactory level.

Unfortunately, as is apparent from the number and intensity of successful attacks, many organizations still have a long way to go before they satisfy even the lowest level of acceptable compliance. In this chapter, we looked at the life cycle for secure system development and indicated which areas need to be addressed. An extensive list of suggested tasks and levels of involvement is provided in Appendix C for those looking for guidance on what to do to improve the security of their mission-critical systems in the face of increasingly sophisticated attacks.

Endnotes

[1] McBride, P., and E. P. Moser, *Secure System Development Life Cycle (SDLC): Building Security Into Your System—Not Bolting It On After the Damage is Done,* Atlanta, GA: Metases Publications, 2000.

[2] DACS (The Data & Analysis Center for Software), *Enhancing the Development Life Cycle to Produce Secure Software, A Reference Guidebook on Software Assurance,* Version 2.0, U.S. Department of Defense, 2008.

[3] Dobbing, B., and S. Lautieri, *SafSec: Integration of Safety & Security Certification, SafSec Methodology: Standard,* Praxis High Integrity Systems, 2006.

[4] Howard, M., and S. Lipner, *The Security Development Lifecycle,* Redmond, WA: Microsoft Press, 2006.

[5] See http://www.mcafee.com/us/resources/data-sheets/foundstone/ds-secure-software-dev-life-cycle.pdf and http://www.slideshare.net/few/basic-of-ssdlc, last accessed on July 20, 2012.

[6] See http://www.prlog.org/11423160-wck-lancelot-ssdlc-security-system-development-lifecycle-offers-support-to-the-sdlc-process.html, last accessed on July 20, 2012.

[7] See http://www.infosecurity-magazine.com/view/23844/was-stolen-symantec-source-code-behind-the-rsa-securid-attacks/, last accessed on July 20, 2012.

[8] The smart grid is essentially the combination of traditional industrial control systems that manage the distribution of electricity, wireless networks, and information systems that collect, analyze, and respond to data obtained from sensors, including intelligent meters, distributed throughout the physical grid. More detail can be obtained from http://energy.gov/oe/technology-development/smart-grid, last accessed on July 20, 2012.

[9]   See http://www.rsa.com/node.aspx?id=3872, last accessed on July 20, 2012.

[10]   See http://www.gdv.ca/files/CA_NT4_Discontinued.pdf , last accessed on July 20, 2012.

[11]   McGraw, G., *Software Security: Building Security In*, Boston, MA: Addison-Wesley, 2006.

[12]   Biggs, C. L., E. G. Birks, and W. Atkins, *Managing the Systems Development Process*, Englewood Cliffs, NJ: Prentice Hall, 1980.

[13]   Federal Aviation Administration, *AQS-200 System Development Life Cycle*, 2007. Available at http://www.google.com/url?sa=t&rct=j&q=faa%20sdlc&source=web&cd=1&sqi=2& ved=0CFkQFjAA&url=http%3A%2F%2Ffaaco.faa.gov%2Fattachments%2FAQS200-001-WI_ProcessDocumentation_Rev1.doc&ei=Cm7CT8mFDqf06AHCnZXGCg&usg =AFQjCNH5p-oa_I9yNEMGCgcC2b5IwXqmWQ. Accessed July 20, 2012.

[14]   Axelrod, C. W., "The Need for Functional Security Testing," *STSC CrossTalk: The Journal of Defense Software Engineering*, March/April 2011, pp. 17–21. Available at http://www. crosstalkonline.org/storage/issue-archives/2011/201103/201103-axelrod.pdf.   Accessed July 20, 2012.

[15]   Calder, A., and S. Watkins, *IT Governance: A Manager's Guide to Data Security and ISO 27001/ISO 27002*, 4th Edition, London: Kogan Page, 2008.

[16]   See http://www.infosectoday.com/Articles/gassp.pdf, last accessed on July 20, 2012.

[17]   See http://all.net/books/standards/GAISP-v30.pdf, last accessed on July 20, 2012.

[18]   See http://www.aicpa.org/InterestAreas/InformationTechnology/Resources/Privacy/ GenerallyAcceptedPrivacyPrinciples/DownloadableDocuments/GAPP_BUS_%20 0909.pdf, last accessed on July 20, 2012.

[19]   See http://www.aicpa.org/InterestAreas/InformationTechnology/Resources/Privacy/ GenerallyAcceptedPrivacyPrinciples/DownloadableDocuments/GAPP_PRAC_%20 0909.pdf, last accessed on July 20, 2012.

[20]   See http://www.first.org/cvss, last accessed on July 20, 2012.

[21]   See http://cwe.mitre.org/, last accessed on July 20, 2012.

[22]   See http://en.wikipedia.org/wiki/Requirement, last accessed on July 20, 2012.

[23]   The firm MetaSes no longer appears to be an existing company; *Secure System Development Life Cycle (SDLC): Building Security Into Your System—Not Bolting It On After the Damage is Done* [1] is unfortunately not available through regular channels. The published book, *Secure Internet Practices: Best Practices for Securing Systems in the Internet and e-Business Age*, does include a section on the topic but refers to [1] for greater detail.

[24]   See http://bsimm.com/, last accessed on July 20, 2012.

10

Safe SSDLC Projects in Greater Detail

*[System safety is] the application of engineering and management prin-
ciples, criteria, and techniques to optimize all aspects of safety within
the constraints of operational effectiveness, time and cost throughout all
phases of the system life cycle.*
—Joint Software System Safety Committee, *Software System
Safety Handbook* [1]

*A secured system ensures that there is no security bug so that security
threats are eliminated, whereas for a safety-critical system, any bug
could be devastating.*
—Asoke K. Talukder and Manish Chaitanya, *Architecting Secure
Software Systems* [2]

Introduction

One might assume that all that needs be done to generate the tasks, subtasks,
purposes, and levels of involvement for safety-critical software systems is to du-
plicate the many tables in Appendix C and substitute the words *safe* and *safety*
for the words *secure* and *security*. One wishes that it were that simple. While,
in many instances, it does make sense to perform such a substitution, there are
many cases where such parallels fall short.

Both the secure and safe software system development life cycle models
have similar, if not the same, heritages. They both derive from systems and
software engineering which are described in Chapters 2 and 3. However, as described in Chapters 5, there are significant differences between the approaches
of those involved in developing secure software systems and those focused on

safe software systems—namely, software security engineers and software safety engineers, respectively.

The main differences are ones of emphasis on specific phases and tasks rather than structural differences. In particular, development life cycles proposed for safety-critical systems pay significantly more attention to the requirements, testing, and certification phases, than do the development life cycles for security-critical systems. This is not to support or encourage this difference—in fact, developers of security-critical systems can learn a great deal from their software safety engineering counterparts, as described in [3].

Definitions and Terms

As we did in Chapter 9 for security-intensive software systems, we must clarify the differences among *safe software systems*, *safety software systems*, and *safe safety software systems*. With respect to safe software systems, safety is a quality *attribute* of the software systems, whereas for safety software systems, safety is the predominant *function* of the software systems. As with the term secure security software system the term safe safety is not redundant, as it might seem to be. This is because the latter term refers to software systems for which the main function is to provide safety services. However, these software systems need to be built so that the safety software systems themselves exhibit strong safety attributes. While difficult to come up with specific examples, one suggestion is to consider a system that locks a door in order to prevent radiation from leaking from a nuclear plant in distress. It is important that a door control system does not fail in a closed default status, since that would result in workers being trapped in the radioactive space. This actually happened at the Fukushima nuclear power station, which was knocked out by an earthquake followed by a tsunami on March 11, 2011. Workers were unable to escape from a highly radioactive space because the doors locked automatically and an operator was unwilling to manually override the system because of fear that the trapped workers would contaminate others [4].

Somewhat more difficult is trying to come up with examples of safe and secure software systems that exhibit both safety and security attributes but have functions other than the provision of safety and security services. Likewise, safety and security software systems are systems with predominant functions in the safety and security spaces. However, we must ask, are the latter both safe and secure? Not necessarily. According to our earlier definitions, such systems are both protective of internally-processed information assets and are designed not to harm humans and the environment. Examples of such software systems are increasing quite rapidly as engineers are combining safety-critical and security-critical software systems (as with the smart grid [5] and a host of other industrial control systems, such as water treatment plants, nuclear power plants, etc). The danger to these control systems was considered to be minimal because they were not connected to any public networks. However, other means of injecting malware, such as via portable USB thumb drives, further negates that argument. The approach to making these systems of systems or cyber-physical systems are both safe and secure is usually based on securing the control systems by ensuring that means of access, of introducing malware, and of effecting denial-of-service attacks, which might be opened up in creating these combined systems, meet high security standards. These standards will often be greater than those that would apply to information systems in isolation because the range and magnitude of the risks faced by systems of systems and cyber-physical systems are raised to much higher levels. This is because the information systems not only must protect data but also ensure that access to the control systems is restricted so as to prevent those with evil intent from gaining control of the authorized access to safety-critical control systems.

Finally, if we talk about safety and security software systems, which are both safe and secure, we mean software systems that function to preserve the safety of humans and security of assets and that also meet safety and security requirements. There are likely not that many examples of such systems, although some of those that do exist have shown particularly bad judgment on the part of decision makers when it came to the inevitable trade-off between safety and security considerations.

Perhaps the primary examples are those software systems that control ingress into and egress from a physical plant. The previously mentioned Fukushima incident is a prime case. Here, the system was designed so that, in the event of radiation being leaked into a room, no one would be able to open the door from the outside without using an interlock system. From one point of view, this is a safety requirement, but it can also be thought of as a security requirement, where the asset (in a negative sense) being secured is radiation. The trade-off, if it was ever considered as such, was between allowing radiation to escape versus risking having workers trapped in the room. In general, safety trumps security because when safety-critical systems malfunction or fail, there is the potential for injury, loss of life, and damage to the environment. However, it often depends on who is making the decisions. Based on the outcome, one might presume that the senior management at TEPCO (the company responsible for operating the Fukushima plant) appears to have put financial returns above the safety of their employees and contractors in their decisions about the level of physical protection. This was somewhat reflected in the heights of walls protecting against tsunamis, although the risk of a tsunami of the size and strength of that which knocked out the plant on March 11, 2011 could have fairly been assumed to be extremely low. Only after the fact did it appear as if TEPCO management had been remiss.

A similar, though not as devastating, situation, which was not reported but came from a reliable source, arose in a building during a Y2K test of the building entry system. A high-security building had an extremely effective system that locked down crash-proof doors and windows. During the test, the system failed and doors and windows closed, preventing anyone from entering or exiting the facility. Engineers were unable to reset the system and open the doors and windows, which were so hardened that rescuers had to bulldoze a hole in a wall in order to gain entry and help the trapped workers to get out of the building. The trade-off here was between protecting the building against intruders and allowing those within the building to escape. It might have been the case that, under normal operating conditions, the system would have allowed the workers to get out of the building, as in the case of a fire for example, but that in its failed state it was not able to release the doors and windows. Clearly, the failure that occurred during Y2K testing had not been anticipated.

The above is summarized in Table 10.1, which is similar to Table 9.1, except that the entries of the table are from the subjective perspective of a system safety engineer rather than a system security engineer.

Table 10.1 shows some awareness by software safety engineers of the need for security, but not as great a commitment to secure attributes as might a software security engineer might have. When it comes to combined security-criticality and safety-criticality, software safety engineers will undoubtedly show a bias in favor of safety aspects.

Hazard Analysis

The fundamental characteristic of safe software systems, as we have discussed, is that such systems should not do harm to humans or the environment as a result of their specified functioning, malfunctioning, or failure. As a result, it is wise to perform a risk analysis of the impact of various hazards to which a safety-critical system might be vulnerable.

An excellent overview of the common approaches to hazard analysis is given in the report from the Joint Software System Safety Committee [6].

Table 10.1
Safety and Security Attributes vs. Safety and Security Functions

Attributes	Functions		
	Safety	**Security**	**Combined Safety and Security**
Secure	Medium	Medium to high	Medium
Safe	Very high	Medium to low	Very high
Safe and Secure	Very high	Medium to high	Very high

According to Military Standard 882D [6], the purpose of which is to establish a System Safety Program (SSP), there are a number of tasks as follows:

- Software requirements hazard analysis;
- Top-level design hazard analysis;
- Detailed design hazard analysis;
- Code-level software hazard analysis;
- Software safety testing;
- Software/user interface analysis;
- Software change hazard analysis.

To this list, we might add: software system decommissioning and disposal hazard analysis.

The Standard [6] is directed at contractors and applies to both system and software requirements, so that it might be more accurate to replace *software* with the words *software system*, based on previous discussions.

The relationship between the various types of hazard analysis and the development lifecycle is illustrated in Figure 10.1.

Software Requirements Hazard Analysis

Examples of software system safety requirements include unsafe modes, such as:

- Out-of-sequence;
- Inappropriate magnitude;
- Inadvertent command;
- Adverse environment;
- Wrong deadlocking;
- Failure to command.

Some of these modes, such as out-of-sequence and wrong deadlocking, are inherently within the software systems themselves; others relate to out-of-range data or calculation results that might arise from a combination of software and data entry errors; others relate to the environment or context within which the software system operates. Some of the above situations can be controlled directly by the system designers and programmers, and others may be out of their control but should be anticipated and accounted for in the design. Operators or support personnel can work around poor design decisions and

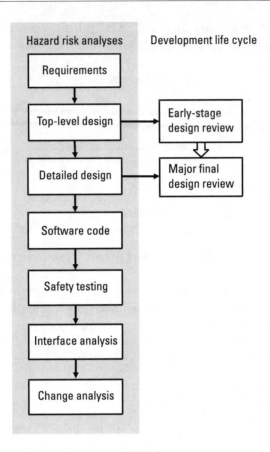

Figure 10.1 Hazard analysis throughout the SSDLC.

faulty program code, which is quite common but not at all a desirable means of dealing with such situations.

Top-Level Design Hazard Analysis

This level of risk analysis covers safety-critical software components and includes the following:

- Definition of safety-critical risk;
- Determination of the degree of risk involved;
- Design to be followed ;
- Creation of a test plan for verifying that the design meets safety requirements.

The results of this analysis should be submitted as input to preliminary design reviews. This type of risk analysis can, and should, be applied to hardware also, particularly since many safety-critical software systems include hardware modules and also control physical systems.

Detailed Design Hazard Analysis

This analysis of the detailed design of the software system should be completed prior to the start of coding software programs. The detailed design takes the results of the hazard analyses for software requirements and top-level design and must ensure that these results are accounted for in the safety-critical software components. The results of this analysis provide input to the final design review.

Code-Level Software Hazard Analysis

This analysis is based on the results of the Detailed Design Hazard Analysis and requires the analysis of program code and system interfaces to determine if there are any errors, faults, or conditions in the code or interfaces that might result in undesirable safety-related events. This analysis should be ongoing throughout the remainder of the life cycle.

Unfortunately, the so-called static analysis of programming code is often an art rather than a science, despite numerous attempts at automating the code review process. The ability of the coding practices to avoid programming hazardous states is also a function of the programming language selected. Some languages, such as Ada, MISRA C, and SPARK, have built-in restrictions that serve to avoid certain unsafe or risky conditions. However, more prevalent languages, such as C, C++, C#, Java, and SQL, are less restrictive and, it can be argued, allow more safety issues, as described in [7].

The consideration of interfaces with other systems is key, as is the need to verify that no new issues are introduced when the systems to which the subject software system is linked and the interfaces between them are changed in any way. Sometimes changes external to the system being analyzed can have major negative impact on the operation of such a system. In addition, the context in which a system operates can change, both within the overall system with respect to platforms and infrastructure, and also with respect to the environment in which the system operates, which could change so as to no longer meet safety requirements.

Software Safety Testing

The purpose of this task is to generate documentation showing that all known hazards have been addressed and that they have been eliminated or reduced to an acceptable level of risk.

There are several issues here that affect all security-critical and safety-critical software systems; namely, that the completeness and quality of the testing depends on how complete the safety requirements are and how extensive the testing is. Safety testing should go beyond checking against the list of known potential safety hazards and investigate the impact of malfunctioning and failure beyond the usual scope. In addition to this nonfunctional safety testing that examines the general safety aspects of the software system, it is important to perform functional safety testing, which looks at hazard conditions that might arise from the software not doing what it is not supposed to be doing.

Another issue relates to who must decide what level of risk is acceptable. An engineer might have a different view of risk from management, for example. In some cases, management overrules the concerns and warnings of engineers because of budgetary and timeliness issues, sometimes with dire consequences.

While many of the methods for reviewing software design, analyzing source code, and testing functionality are common to both security-critical and safety-critical software systems, safety-critical control systems use methods specific to control software systems, such as model checking, abstract interpretation, and theorem proving as described in Feron [8].

Software/User Interface Analysis

The objective of this task is to ensure that software user procedures are fully developed and documented.

Human error remains one of the main reasons for software system malfunctioning and failure, and such errors can be greatly reduced by proper training in the use of the system, as well as by responding to feedback from users with respect to observed ambiguities, incorrect functioning, unexpected system responses, unintelligible error codes, and the like. While human beings can compensate for many system deficiencies, their ability to work around and compensate for incorrect or hazardous system operation is limited by their understanding of how the system is supposed to work. Also, many modern technologies respond in much quicker and more complex ways than human operators, meaning that compensation for system misbehaviors is not always possible, and is particularly lacking if the software system is under a cyber attack or is functioning in inexplicable ways because of errors in the software or unanticipated user actions.

An example of errors causing wrong responses is the crash of Air France Flight 447, which is documented in a report that describes how the Air France Airbus 330 crashed into the ocean due to a "slow speed stall" that likely resulted from the pilots not being able to properly take over manual controls when the fully-automated aircraft-control system failed [9]. The failure of the automated system aborted because all the primary and backup airspeed sensors failed due to icing. A possible contributing factor was that the Airbus 330 uses a fly-by -wire system, which takes over virtually all pilot functions. As a result, the pilots were thought not to have had enough experience with manually controlling the aircraft.

When software-based control systems are particularly safety-critical, it may not be adequate to run the software through a series of tests and then rely on users to point out discrepancies between what was required and what was manufactured. In such cases, it is helpful, if not mandatory, to prototype the user interface and build simulation models on which users can be trained to handle many abnormal situations without them being exposed to the consequences of actual crashes.

Software Change Hazard Analysis

When software systems, their platforms and infrastructures, or their operational contexts are changed, there can be considerable impact on potential safety hazards. Not only must all prospective changes, modifications, patches, and the like be analyzed prior to making the change to the greatest extent possible, but there needs to be a tracking and monitoring program in place to ensure that important changes affecting safety have been made and continue to be in force.

Sometimes a fix can make the situation worse, or a problem can occur in a seemingly unrelated part of the system. It is no wonder that making security and software changes once a system is operational can be orders of magnitude more costly than if the problem had been resolved in the early stages of the development life cycle.

In a certain respect, the decommissioning and disposal of a safety-critical system can be considered to be in the major change category. Not only do the hazards resulting from changes need to be considered—here in terms of replacing the existing system with a new system, but the decommissioning process itself also heightens risk from users and operation/support staff having to maintain both the former and replacement systems in parallel, having to learn new systems and procedures, and the like.

In addition, there are potential hazards in the disposal of software-based systems. For example, the system might be taken up inadvertently or purposely by an unauthorized party and misused because of lack of training and support, causing risk to the new users. Or the system might be used against those who originally developed the system, as might be the case for a weapons system that has been discarded. This suggests that oversight of the decommissioning and disposal processes is very important from both the safety and security perspectives.

It should be noted that the decommissioning and disposal phases are often not given the attention that they deserve and need. This is further exacerbated by auditors and others with oversight responsibility wash their hands of a system once it has become obsolete and not monitoring how the systems are discontinued.

The Safe Software System Development Lifecycle

As with the details of the secure SSDLC, shown in the tables of Appendix C, another set of tables, this time addressing the safe SSDLC, is given in Appendix D. Noteworthy differences between the entries in the tables of Appendix C and Appendix D include the additional focus on hazard analysis, verification, and validation in the latter appendix. While these functions also play a part in the design, development, and testing of security-critical software systems, they are not given nearly the attention that they get with safety-critical software systems. Furthermore, there is usually more attention paid to the choice of programming languages, platforms, and infrastructure when the safety of software systems is at stake.

In Figure 10.2 (which is similar to Figure 9.2 except that it refers to safety instead of security), we can see how various safety aspects affect the individual phases of the SSDLC. Noteworthy differences include a greater emphasis in the safety case on risk analyses in the early stages of the development life cycle, as well as a lack of penetration testing during the implementation, deployment, and operational phases. This is in keeping with the cultural and technical perspectives of software safety engineers.

In Figure 10.3, we show the hazard risk management process used in the design of safety-critical systems mapped against equivalent security risk factors. It is immediately apparent that the hazard risk analysis is a much more refined and extensive system that the comparable security risk analysis.

We will now briefly elaborate on the steps in the hazard risk management process as follows:

- *Identify potential hazards:* if this were readily achievable at reasonable cost, then one would not be surprised as frequently by unusual events. However, it is clear that even some high impact hazards are missed in part because identifying hazards is bounded by the ability of analysts to anticipate highly unlikely events, especially if they have not occurred previously (see item on *identifying new hazards*). This effort is also limited by time and resources available.

- *Estimate severity:* for safety-critical systems, as previously discussed, hazards are categorized by their consequences which can be catastrophic, critical, marginal, or negligible. Security systems' consequences are often

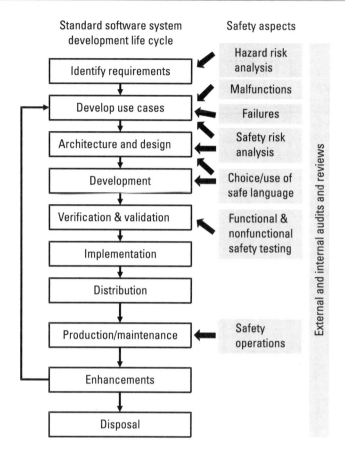

Figure 10.2 Safety aspects related to the SSDLC.

critical but seldom seen as catastrophic, even when they are, because there may be little visible impact.

- *Identify causes:* for some safety-related incidents, the causes are easily identified, but in many cases, such as plane crashes, accurate causes are often not determined until data from the flight recorders, are analyzed. Consequently, anticipated relationships between causes and effects are difficult to develop.

- *Estimate likelihoods:* as we discussed in Chapter 6, it is very difficult to estimate the probability that a particular adverse event will happen, especially for low-probability, high-impact events. Nevertheless, it is a worthwhile exercise to attempt to derive some probability estimates because, even if they are off the mark, they might still help analysts and decision-makers to put certain risks in perspective.

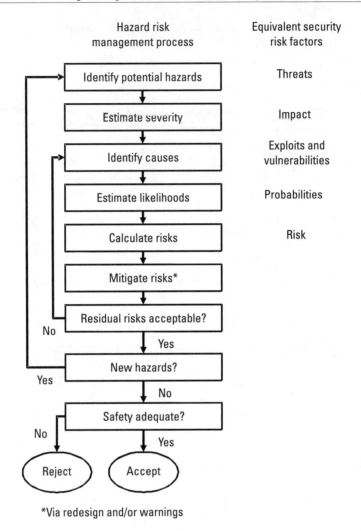

Figure 10.3   Hazard risk management process and the equivalent security risk factors.

- *Calculate risks:* as we also discussed in Chapter 5, there are many different ways in which to calculate risk, many of which are erroneous or do not contribute to decision-making. With safety-critical systems, such as avionic and nuclear plant control systems, it is worth investing a considerable amount in order to generate useful results on the likelihood of system malfunctions and failures.

- *Mitigate risks:* while the common view of risk mitigation for security-critical, software systems is to make sure that coding is correct and that there are protective tools in place, this is not sufficient for safety-critical systems that are required to give sufficient warnings of imminent failures. It may be necessary to redesign and rewrite the software to avoid such risks. There is usually, but not always, more of an interest in mitigation for safety-critical systems than for security-critical systems.

• *Determine if residual risk is acceptable:* the main issue here is whether the risk that remains after various mitigation strategies have been applied is acceptable to decision-makers. The problem here, as has been discussed, is that risk acceptance is highly dependent on personal subjective feelings. Also, the degree to which a risk can be transferred to someone or some entity has a lot to do with the decisions.

• *Determine new hazards:* the hazard environment is changing rapidly, and so even a recent risk analysis could be out-of-date because of situational changes of many different kinds. It is important to continually update the risk analysis, even when the system has already been deployed.

• *Determine whether safety is adequate:* here we have the same problem as for risk assessment; whether the safety is enough is somewhat subjective and varies significantly with who is making the decisions and what their agenda might be.

This is clearly a process from which those designing security-critical software systems could benefit and one that becomes mandatory when one combines separate systems into complex cyber-physical systems.

Combined Safety and Security Requirements

The main thesis of this book is that safety and security engineering for software systems, while deriving from the same evolutionary tree, have become separate branches, even though the trend is for safety-critical and security-critical software systems to be functionally combined into systems of systems with both safety and security attributes.

One of the few references to address combined safety-critical and security-critical software systems certification is SafSec. In this regard, Lautieri [10] makes the following statement:

> Many systems, particularly in the military domain, must be certified or accredited by both safety and security authorities. Current practice argues safety and security accreditations separately. A research project called SafSec has been investigating a combined approach to safety and security argumentation, and has shown that there can be practical benefits in performing a combined analysis and documenting a combined argument for both safety and security.

SafSec Methodology: Standard [11] and the *SafSec Methodology: Guidance Material* [12] are available for download from the Altran Praxis website. *SafSec Methodology: Standard* [11] includes a good set of definitions of terms used. It is apparent from the SafSec documents that security and safety are considered to be contributors to a system's dependability.

This same definition of dependability as having security and safety components is used by Firesmith [13]. While such a definition is reasonable, the concept of dependability is considered to be too broad for detailed examination, which is why we treated safety and security as separate, specific attributes of software systems. Indeed, dependability includes other important attributes relating to reliability, resiliency and survivability, but these represent areas of study that require dedicated research and exposition. Consequently, we shall leave these latter issues for others to follow up on.

Summary and Conclusions

We have seen in this chapter that those who design and develop safety-critical software systems have a very different perspective in regard to the various phases of the SSDLC.

Software safety engineers focus on the risks of their systems doing harm if they were to fail or malfunction and are less concerned about someone getting into their systems and stealing sensitive information or gaining control of the system. However, as we see increasing numbers of complex cyber-physical systems, it is clear that these same software safety engineers need to be aware of the dangers of attacks from outside or from within, since it is not just a matter of protecting information, but also covers concerns about an attacker gaining control of safety-critical systems and forcing it to malfunction, fail, or behave in undesirable and dangerous ways.

The viewpoints of software security engineers and software safety engineers towards the design and development of software systems are so divergent that one might question whether these individuals, upon whom our lives and livelihoods so much depend, will ever collaborate to the degree necessary to ensure that modern systems of systems meet prudent security and safety requirements. Even those who are promoting collaboration between software safety and security engineers, such as the proponents of SafSec, are not hopeful that such partnerships will ever come together. However, such commonality of purpose is inevitable, except that it is likely to take many more years to come to fruition.

It is only through continuously driving home the point that safety and security are different animals but that will have to learn to live together in harmony if there is to be any hope of reaching an acceptable level of safety and security in future software systems.

Endnotes

[1] Joint Software System Safety Committee, *Software System Safety Handbook: Technical & Management Team Approach*, 1999. Available at http://www.system-safety.org/Documents/Software_System_Safety_Handbook.pdf. Accessed July 23, 2012.

[2] Talukder, A. K., and M. Chaitanya, *Architecting Secure Software Systems*, Boca Raton, FL: Auerbach Publications, 2009.

[3] Axelrod, C. W. "Applying Lessons from Safety-Critical Systems to Security-Critical Software," *2011 IEEE LISAT (Long Island Systems, Applications and Technology) Conference*, Farmingdale, NY, May 2011. Article available for purchase at http://ieeexplore.ieee.org/xpl/login.jsp?tp=&arnumber=5784222&url=http%3A%2F%2Fieeexplore.ieee.org%2Fiel5%2F5775772%2F5784200%2F05784222.pdf%3Farnumber%3D5784222. Accessed July 23, 2012.

[4] See http://in.reuters.com/article/2011/03/25/idINIndia-55893620110325, last accessed on July 23, 2012.

[5] The smart grid was described in some detail in Chapter 11.

[6] U.S. Department of Defense, *MIL-STD 882D, System Safety Program Requirements*, 2000. Available at http://www.atd.dae.mi.th/download/mil-std-882d%5B1%5D.pdf. Accessed July 23, 2012.

[7] Axelrod, C. W., "Trading Security and Safety Risks within Systems of Systems," *INCOSE Insight*, Vol. 14, No. 2, July 2011, pp. 26–29.

[8] Feron, E., "From Control Systems to Control Software," *IEEE Control Systems Magazine*, Vol. 30, No. 6, 2010, pp. 51–71.

[9] Clark, N., "Report on '09 air France Crash Cites Conflicting Data in Cockpit," *The New York Times*, 2012. Available at http://www.nytimes.com/2012/07/06/world/europe/air-france-flight-447-report-cites-confusion-in-cockpit.html, last accessed on July 23, 2012.

[10] Lautieri, S., D. Cooper, and D. Jackson, "SafSec: Commonalities between Safety and Security Assurance," *Proceedings of the Thirteenth Safety Critical Systems Symposium*, Springer-Verlag, 2005.

[11] Dobbing. B., and S. Lautieri, *SafSec: Integration of Safety & Security Certification—SafSec Methodology: Standard, S.P1199.50.2, Issue 3.1*, Praxis, 2006. Available via a link at http://www.altran-praxis.com/safSecStandards.aspx. Accessed July 23, 2012.

[12] Dobbing. B., and S. Lautieri, *SafSec: Integration of Safety & Security Certification - SafSec Methodology: Guidance Material, S.P1199.50.3, Issue 3.1*, Praxis, 2006. Available via a link at http://www.altran-praxis.com/safSecStandards.aspx. Accessed July 23, 2012.

[13] Firesmith, D. G., Common Concepts Underlying Safety, Security, and Survivability Engineering, Technical Note CMU/SEI-2003-N-033, Software Engineering Institute, Carnegie Mellon University, 2003.

11

The Economics of Software Systems' Safety and Security

Security is an area where taking shortcuts can lead to disaster.
—Eric Allman [1] Multifaceted software engineer

Introduction

The question asked in this final chapter is if safety and security will ever be adequately incorporated into software systems and whether such safety and security improvements can be economically justified now and in the future. The purpose of this analysis is to determine how one might create safer and more secure software systems using justifications based on behavioral economics rather than on some general sense that it is a good thing to do.

In prior chapters, we examined the origins of systems and software engineering, enumerating specific requirements that need to be met in order to try to ensure that safety and security will be built into software-intensive systems. We also looked at the extent to which those requirements are, and are not, being met in practice. Having gone through this entire process, there remains a sense that perhaps we will never completely master these needs and produce completely protective and protected software systems. The reasons for such shortfalls are many, including lack of knowledge and training of those given responsibility for safety and/or security aspects of software-intensive systems and their having neither the skills nor inclination to accept that responsibility. Another possible reason is the increasing cost burden demanded by complex integrated software systems and cyber-physical systems that operate in ever-higher risk environments, and yet for which sufficient monies are not made available to create essential security and safety.

Modern software systems are victims of rapidly increasing numbers of attacks and of attackers acquiring greater sophistication and effectiveness. Also, a proliferation of unintentional errors and uncontrollable external and internal events often lead to even greater adverse financial consequences. There are clear indications that the gap is rapidly widening between the exposure of these systems to malware, malfunctions, and failures in comparison to the amount and effectiveness of the efforts to bring existing and new software systems to a reasonable level of security and safety.

Put another way, there is clearly a growing divergence between the "supply" of threats to the safety and security of software systems along with successful exploits than result from those threats, and the "demand" for huge investments in safety and especially security, which are not keeping up with the risks.

Closing the Gap

What can be done to change these odds in favor of those trying so hard to infuse software systems with appropriate levels of safety and security? How important is it to correct the huge number of severe weaknesses still inherent in legacy systems? What is needed to ensure that future software systems can be trusted to be secure and safe in the face of countermanding trends? These are questions that are frequently asked, and to which the answers are mostly disappointing. Perhaps we are asking the wrong questions or those hearing the questions do not fully understand them. After all, where the pressure for improving safety and security is most intense, software engineers have responded with respectable attempts to shore up those systems, so we know that it can be done, albeit at high cost.

Or perhaps there are underlying psychological reasons. Beth Gardiner [2] presented the results of studies by psychologist Robert Gifford of the University of Victoria, British Columbia, who determined "behavioral barriers to combating climate change." Gifford presented the following four behavioral characteristics, which could just as easily be applied to cybersecurity:

- We have trouble imagining a future drastically different from the present;
- We block out complex problems that lack simple solutions;
- We dislike delayed benefits and are therefore reluctant to sacrifice today for future gains;
- We find it harder to confront problems that creep up on us than emergencies that hit quickly.

In the 1960s, typical business-oriented computer systems would fail on a daily basis. A common remark at that time was that if airplanes were as unreliable as computers, nobody would be willing to fly. Since then, reliability of hardware components of computers has improved by orders of magnitude. It might therefore be construed that software components are also much more reliable. However, the complexity and diversity of software systems and their aggregation into so much more complicated systems of systems have worked against software developers in their attempts to eliminate security vulnerabilities and to keep occurrences of malfunction and failure as low as possible. This is why system architects have heretofore pushed to isolate safety-critical systems from public communications networks since avoidance is pretty much the only approach to use when protection and prevention are not viable. However, requirements for increased functionality on the one hand, and reduced costs on the other hand, have pushed engineers into realms that are uncomfortable, unsafe, and not secure.

The fundamental question is whether there is any real motivation to make software systems safer and more secure or whether the diminishing returns of increasing safety and security mean that, as time progresses, stakeholders are less inclined to embark on the necessary activities to improve the safety and security attributes of software-intensive systems. As supported by the Gifford study mentioned above, human nature may be such that people will only respond to disasters, particularly with respect to cybersecurity, and that it will take the occurrence of a major cyber or cyber-physical attack before government and the private sector will respond to previously ignored warnings. For example, Vice Admiral Michael McConnell (USN, Ret.), in his February 23, 2010 testimony before the U.S. Senate Committee on Commerce, Science, and Transportation, said that risks from cyber attacks will not be mitigated until there is more active government involvement and that such involvement will not be forthcoming until a "catastrophic event" actually happens [3].

Just as the public's unwillingness to accept frequent airplane crashes has led to much more stringent standards and certifications in the airline industry, such as DO-178B, disastrous or catastrophic cyber events may be the only real impetus for improving security.

However, there is another view; namely, that a greater understanding of the forces behind motivations of various participants might result in actions that, by assigning safety and security as personal risks to various players, actions will result that will mitigate those risks. But in order to accomplish this, it is first necessary to identify, understand, and quantify how particular risks affect individuals and the degree to which various players can indeed effect change.

In this chapter we will look at the following:

• Technical debt in regard to the development of software systems;

- The impact of security and safety on individuals and groups according to their value functions;

- The means of persuading or forcing such individuals and groups to take actions to minimize their personal exposure and by so doing, optimizing globally;

- An agenda for optimizing the objectives of all those involved in the creation, operation, and use of safety-critical and security-critical software systems.

The goal is to design a program that over time will reduce exposure of security-critical and safety-critical software systems to an acceptable level, despite the fact that the bar is continually being raised.

We will discuss ways to detect, protect against, and fight tampering as a means of responding to increased threats in the cyber and physical worlds. Finally, we will take a look at the use of "security and safety patterns."

Technical Debt

The concept behind technical debt is that, as software systems are designed, developed, implemented, operated, and replaced, those involved in the development life cycle and the subsequent operational environment are taking on, knowingly or not, technical debt that can be either borne or reduced, consciously or otherwise.

Allman [1] states that "... technical debt is acquired when engineers take shortcuts that fall short of best practices." Of course, so-called "best practices" might also fall short of ideal practices, which implies that there are in fact two or more layers of technical debt in that common practices often exclude essential elements, depending upon the authority and vision of the sources of these practices.

Allman gives a number of examples where technical debt might be taken on due to competition between best practices and other factors. The following are some of his examples:

- Setting too-tight deadlines;

- Not anticipating the costs of tools and using them inappropriately;

- Assigning engineers with inadequate or inappropriate skills to projects;

- Skipping documentation requirements or writing inadequate documentation;

- Skimping on good security practices;

- Using obscure and incomplete error messages;

- Using slow, simple algorithms rather than better algorithms that are complex and run more efficiently.

There are clearly a host of other factors to be considered such as under-budgeting projects, using inexperienced project managers, and the like.

All of these deficiencies lead to the amassing of technical debt. As pointed out by Allman, various constituencies assume technical debt according to their roles and their power to limit that debt. Also, they can accrue and transfer technical debt to others knowingly or inadvertently, much as we described in Chapter 6. This is a form of moral hazard where those who create the debt manage to avoid blame or responsibility for controlling and repaying the debt.

Allman provides a good exposition of how technical debt impinges on various areas within and outside organizations. His lists of characteristics and shortfalls are supplemented with items inferred from those lists. Augmented lists are provided in Table 11.1.

While the lists in Table 11.1 are not exhaustive, they do contain helpful indicators as to what is important to various participants in the development life cycle from their particular viewpoints. Desired characteristics and typical shortfalls clearly vary greatly between and among participants depending on their personal agendas, organizational roles, and vulnerability with respect to having technical debt assigned to them. Suggesting how participants might react to factors that affect them personally goes a long way towards helping to determine how they might be influenced to act in a manner that optimizes the overall system rather than just the specific individual components for which the individuals are responsible. Such methods as pricing, taxing, and other discouraging activities, and incentives, such as bonus payments for good performance in the areas of security and safety, can be used to encourage individuals to work in the interests of all stakeholders rather than in his or her personal interest.

Application of Technical Debt Concept to Security and Safety

Allman [1] does reference security as an issue with respect to technical debt. However, his treatment of security is somewhat marginal and there is no mention at all of safety in the article. This is not unusual since most researchers in software development have backgrounds in information or cyber systems rather than industrial control systems. However, the amassing of technical debt is very pertinent to security and, to a somewhat lesser extent, to safety, because of the permeating of risk analysis throughout the SafeSSDLC. Of particular note is the previously mentioned aspect that as software development progresses it becomes increasingly costly to "bolt on" security and safety functionality. At

Table 11.1

Augmented Requirements and Shortfalls from Various Perspectives

Perspectives	Desired Characteristics	Typical Shortfalls
Customers	Software systems that: · Work (operate as expected) · Are understandable · Can be [readily] maintained over the long term · Are reliable [and resilient] · Incorporate industry-standard security [and safety] practices · Can be extended (i.e., are scalable) · Can be enhanced (i.e., modified in response to new requirements) · Are easy and enjoyable to use	Inadequate user documentation Unstable software Frequent malfunctions and failures Unexpected and inaccurate results from specific user actions Inaccurate processing and results Inability for customers to initiate corrections, changes, and enhancements Unavailable, unresponsive technical and operational support staff Lack of knowledge of technical and operational support staff
Help Desk	Well-designed interfaces for the user support function Extensive documentation of features and functions of the software systems Frequently asked questions (FAQs) that users might view prior to requesting to speak to help desk staff Direct and immediate access to knowledgeable technical support	Poorly designed interfaces Bad or nonexistent documentation Indirect effect of code obscurity Long time to respond to customer questions No direct access to those able to solve problems
Operations	Software products that are: · Reliable · Maintainable · Understood DevOps approach whereby operations staff work with development staff at early stages of the life cycle Ready access to the code Availability of up-to-date, accurate, and complete documentation	Poorly designed interfaces Bad or nonexistent documentation Limited access to those able to solve problems Limited access to other resources, such as code, procedures, etc.
Engineers	Developers: Detailed, accurate, unambiguous, complete specifications Agile methods allowing for user sign-off on design at early stages Maintainers: Well-documented code and operational procedures Easy-to-maintain code Easy-to-change procedures without jeopardizing functionality or ease-of-use	Developers: Ambiguous and incomplete specifications Absence of suitable test cases Inadequate knowledge and training, especially relating to security, safety, performance, resiliency, etc. Maintainers: Undocumented or minimally documented code Inadequate, inaccurate, and out-of-date operational procedures
Management	Understands risk management and balances out demands of internal departments, business partners, and customers Understands technical debt and how to minimize it across the lifetime of the product	Favors one area over another to the detriment of the organization Does not understand concept of technical debt Incurs considerable amounts of technical debt without knowing that such debt has been amassed

Specific additions to some of the items in Allman's existing list of characteristics are shown in square brackets.

some point, it becomes impossible to justify even minor corrections, patches, or additions. The result is that software systems frequently remain vulnerable to attack, malfunction, and failure because no one could present a reasonable and convincing case in favor of assuring that certain security and safety features are implemented.

The transfer or avoidance of technical debt is particularly prevalent in cybersecurity as there are huge swaths of systems and networks for which no one is willing or able to take on the requisite responsibility and liability. Nor would they be able to assume the technical debt even if they were willing to do so. The "tragedy of the commons," in which the securing of common resources is not assigned to anyone, and "moral hazard," whereby liability is avoided by those directly involved and consequently transferred to others, such as the general public. Both the tragedy of the commons and moral hazard work against the assignment of technical debt to its rightful owners.

Conversely, when it comes to safety, there appear to be relatively clear lines of responsibility, except when the event is so large, as in the March 11, 2011 tsunami that hit the northeastern coastline of Japan and destroyed the Fukushima nuclear power plant, that no single entity, in this case Tokyo Electric Power Company (TEPCO), which owns and operates the Fukushima facility, is able to afford to assume full liability. Therefore it becames a partial government responsibility to deal with the consequences. Interestingly, when TEPCO issued its final report on June 19, 2012, they agreed that they had not been sufficiently anticipated the catastrophe and that they were not adequately prepared for a tsunami of the magnitude experienced [4].

System Obsolescence and Replacement

Allman [1] seems to suggest that when one system is expected to be replaced by another, the technical debt, which has been accrued by the original system, is somehow forgiven and disappears. This is not necessarily the case, especially for systems that are not retired within the anticipated time frame. A prime example of this was Y2K. Billions of lines of code were written decades prior to the year 2000 with the expectation that they would be replaced by the time the end of the century rolled around. That assumes that the problem emanating from representing years with two digits, rather than four, had even occurred to the programmers, which it likely did not. The technical debt on this oversight ran into several hundreds of billions of dollars. Similarly, when Microsoft decided that it would replace an obsolete operating system NT 4.0, it had to deal with stragglers, who were not ready to upgrade their systems. Customer pressure can mean that vendors have to provide support for software products well beyond the date on which they are originally intended to abandon the product.

The Responsibility for Safety and Security by Individuals and Groups

We now consider how the safety and security of software systems, in particular, are viewed by a wider-ranging constituency and the extent to which safety and security attributes are really considered to be the responsibility of various groups. In a sense, this approach is similar to technical debt in that the cost of neglecting security and safety is accrued throughout the software system development life cycle. Here we look at the economic consequences of security lapses and safety failures on a boarder population comprising those directly and indirectly involved in the creation, deployment, and operation of software systems.

Basic Idea

The fundamental concept behind the proposed risk-based value approach is that stakeholders have a personal view of how including or excluding particular safety and security software features will directly affect them as individuals. While there may be other motivations, an individual will still view the impact of safety and security features and capabilities from his or her personal perspective. Individual motivations will likely only coincide with organizational goals if personal benefits result from a healthy and prosperous organization. This might appear somewhat cynical, but it is not meant as a criticism and does not deny that people will often act altruistically for the benefit of a group, organization, or country. The point is that one's assessment and response to risk is generally highly subjective and intensely personal.

As an example, if a group of developers is responsible for completing an application by a certain deadline and within a predetermined budget, and achieving these goals is the basis for team members being rewarded, then any time delays and additional costs or resource requirements that might threaten the official deadline will be avoided as much as possible. It is only if criteria for rewards and benefits are included in safe and secure software design, architecture, development, and testing that the proper amount of attention will be given to these areas. However, the decision to tailor compensation and other rewards to how well particular safety and security requirements are met has to come from high-level management within an organization.

By recognizing the motivators for security-related decisions and incorporating them into an economic model of how individuals respond to motivators, one understands why certain decisions are made for incorporating security and safety into software development, operation, and decommissioning of software-intensive systems. To quote from Axelrod [5]:

> One cannot develop effective economic models for information security and privacy without having a good understanding of the motivations, disincentives, and other influencing factors affecting the behavior of

criminals, victims, defenders, product and service providers, lawmakers, law enforcement, and other interested parties Predicting stakeholders' actions and reactions will be more effective if one has a realistic representation of how each of the various parties will respond to internal motivators and external stimuli.

As we have seen, safety engineering is very formal in certain critical industries, such as aerospace, avionics, and vehicular and power plant safety, requiring that systems be evaluated based on their potential for doing harm, which then points to a level of certification needed by the industry.

Extending the Model

The model in [5] discusses the motivations by various stakeholders to attack and defend security-critical software systems. Here, we extend the model to cover safety-critical software systems and focus the impact of stakeholder motivations on the SSDLC.

We now examine each of the phases of the life cycle and see how stakeholders are currently involved with or influence each stage. We then recommend how we might change the list of players involved at each stage and their roles and responsibilities in order to achieve a more acceptable level of safety and security. We also suggest mechanisms, both motivators and deterrents, which might be used to influence the outcomes of each step so that the final product or service meets an organization's safety and security requirements. The intent is to come up with a list of actions that will move decision makers and stakeholders towards security and safety postures that best meet the needs of society as a whole. While many of the suggestions will be normative, and will be unlikely to be put into effect because of factors such as high costs, limited resources, and lack of authority, the model at very least points to areas that need attention, and to some extent suggests new and different ways to improve the situation as it stands today.

Concept and Requirements Phase

When requirements for a system are being determined, those making decisions as to what the functional and nonfunctional requirements of the proposed system might be are unlikely to consider a full range of security and privacy issues. The likelihood that safety issues will be included is usually much higher, in large part because safety-related software-system malfunctions and failures might result in injury or loss of life. Insurance companies have formulae for determining the value of an injury or death in monetary terms for their purposes, but often the indirect impact is much greater in terms of loss of confidence in the manufacturer of a defective system, additional burdensome regulatory and audit requirements, and loss of business.

One needs to include a full range of representatives of key areas after a concept is originated and when in the requirements development phase. Concept origination may be a result of ideas from internal business and technical staff, or senior management and/or marketing may become aware of new technologies and competitors' systems, legal and regulatory requirements, obsolescence of current systems, and so on. The concept then needs to be described in terms of the requirements of sponsors, users, and technical, and operational staff, where each participant provides inputs to what then becomes a set of functional, and hopefully nonfunctional, requirements.

In Table 11.2, we see a breakdown of the SSDLC into its component phases and indications as to which stakeholders are involved in each phase and what their roles should be. For the origination and requirements phase, there is usually a mix of business, technical, and operational participants, each of whom brings his or her own perspective to the formulation of the needs of the system. Thus, senior business management and marketing will be most interested in ensuring that the software system provides needed functionality. They also want to be assured that developing the system will take a specific amount of resources and money within a time frame that may be determined by external factors such as time to market (or time to value). The technical staff will want to ensure that currently available technologies can deliver the desired functionality at a reasonable cost within the time parameters. The value profiles of clients, such as management, marketing as a customer surrogate, and end users, will push to maximize functionality and minimize cost and time to market, whereas the personal risks of technology and operational teams will cause them to move in the opposite direction of reduced functionality, increased resources, and more time. A balance between the two groups needs to be agreed upon and is often determined by the relative power and authority of those involved in the negotiations. The client side, consisting of senior corporate and business unit management, often prevails, and the technical and operational development teams are forced to concede. This may be much of the reason behind so many deficient and defective software systems.

The list of areas that are often excluded from this phase of the SSDLC is disconcertingly long, particularly as deficiencies in requirements, particularly the nonfunctional requirements, could likely be avoided if the listed areas had been included in the discussions and had a vote in the decision-making process. Occasionally a senior executive will make a pronouncement about one of these neglected areas having to be included in the process and having his or her direct support, as was the case when Bill Gates of Microsoft mandated building trustworthy secure software. However, much of the time, it appears that many such areas are excluded to the detriment of the project and with the result that these

Table 11.2

Participants and Decision-Makers for Each SSDLC Case

SSDLC Phase	Typical Participants	Usual Decision Makers	Suggested Additional Participants (Often Excluded)
Origination and requirements	Senior management (if strategic system) Business unit management Business unit operations Marketing (if appropriate) Internal-user representatives Information technology Safety engineering (if safety-critical system)	Senior management (if strategic system) Business unit management Information technology management Safety engineering (if safety-critical system)	Information security Security administration Internal auditors Risk management Software assurance Application security analysts Application safety analysts Internal user representatives Customer user representatives Computer operations Technical support Help desk Software certification liaison Identity and access management Legal department (representing lawmakers and law enforcement) Compliance department (representing regulators and external auditors) System administration Business continuity planning Disaster recovery planning
Design, architecture, and specifications	Systems analysts Software architects	Systems analysis management Software architect management	Information security Internal auditors Application security analysts Application safety analysts Computer operations Technical support System administration Business continuity planning Disaster recovery planning
Development	Developers Technical support	Development management	Information security Application security analysts Application safety analysts Business continuity planning Disaster recovery planning
Verification (reviews, code inspections)	Business unit sponsors Systems analysts	Systems analysts	Application security analysts Application safety analysts
Validation (testing)	Business unit sponsors Developers Testers	Testers	Application security analysts Application safety analysts
Deployment, operations, maintenance, and technical support	Customer relations Marketing Operations Network engineers Technical support Field service engineers	Marketing Operations	Information security System safety Systems analysts Business continuity planning Disaster recovery planning
Decommissioning and disposal	Business unit sponsors Computer operations	Marketing Computer operations	Information security System safety Internal and/or external auditors

areas, which are usually called in during the later stages of development and testing, if at all, are forced into the predicament of either accepting substandard systems or trying to bolt on security and safety, achieving better performance by adding processing capacity, writing contracts that protect against inevitable malfunctioning, and so on. As was discussed previously, the costs of these after-the-fact fixes can be orders of magnitude greater than if the need for security, safety, resiliency, and the like had been incorporated at the requirements phase. Since there are usually little or no additional available funds during the later stages of the life cycle, the compromise is often to do some minimal touch-up and force people to work around the issues. Perhaps the one exception to this general situation is when the software systems are safety-critical and need to be certified according to very specific safety requirements, in which case safety engineering participants are included. However, this does not necessarily mean that appropriate *software* safety engineers get involved at this stage.

By not including the additional participants list in the right-hand column of Table 11.2, the cost to the project will inevitably be greater. If these participants are heeded later in the project, the some of their requirements will be addressed. If not, then the costs will be in the form of opportunity costs or indirect costs that may not be linked back to the root cause of the deficiencies. For example, if security is not built into the software system, then expensive tools and personnel will need to be added when the system is deployed in order to provide some higher level of security. However, application firewalls, intrusion detection and prevention, and similar technologies do not make up for vulnerabilities inherent in the software system and are a continuing cost drain for their deployment, maintenance, and management.

From a value perspective, we see that staff with power and authority pushes to minimize what they believe to be costs attributable directly to them. They will try to maximize the benefits (other cost savings, competitive edge, etc.), and ignore the remonstrations of those who are essentially tasked with controlling the risks related to the software systems. Unfortunately, over time, as the costs of early-stage neglect mount up, it is often those who were not consulted or heeded in the initial phases of the SSDLC who shoulder the blame and bear the brunt of the negative consequences. Ultimately, someone has to pay for the inadequacies of the requirements phase—and often as not it is those who were excluded from the process. The overall aggregated costs are easily as high, if not a great deal higher, than they would have been had the right type and level of resources been brought to bear earlier in the life cycle.

Design and Architecture Phase

For the most part, the design of software systems and their architecture (platforms, networks, etc.) is left to technicians. This is because this phase usually consists of taking a set of requirements, as developed in the prior phase, and translating those requirements into specific designs comprising application modules, system software, physical systems, and networks.

There are many problems related to this phase of the process. First, it relies on the quality and completeness of the requirements and also on the knowledge and experience of the software-system designers and architects, any of which could be lacking. Second, designers and architects are motivated to come up with designs that will meet requirements, whether or not requirements are adequate. Thus, for example, nonfunctional attributes, such as security and performance, may not be adequately covered in the requirements and hence will likely not be accounted for in the design. In many cases performance, particularly for systems using innovative technologies, cannot be predicted unless a prototype is developed or a simulation is built. Therefore, in order to motivate designers and architects to develop designs that will perform satisfactorily, or prevent them from suggesting architectures that will not meet performance goals, one needs to appreciate the importance of the attributes to those for whom the value of performance, security, resiliency, data integrity, and so on is high. Then one has to ensure that appropriate measures are taken to encourage full consideration of these attributes.

One aspect that is often neglected is the need to incorporate the necessary monitoring, data collection, and analysis, particularly so that users of the system can be tracked. This type of information is critical for computer forensics following a data breach, for example, and yet the need for it to be included as requirements is seldom recognized. However, when the need is expressed, it is usually received negatively as it will lead to immediate increases in costs and resources, along with some operating overhead, and the benefits will not show up for months or years, if at all.

For these reasons, there needs to be oversight from security and safety software engineers, auditors, support staff, and so on in the design stage to make sure that their interests are covered and to verify that the design correctly interprets requirements . This is shown in Table 11.2 in the last column of the design/architecture row.

Development

During the development phase, programmers and other technical staff are generally focused on writing code that is functionally correct and error-free and completed within the assigned timeframe. As with design teams, developers are not required to be particularly creative, except in writing efficient and effective programs. They are usually rewarded for following requirements to the letter rather than trying to second-guess them. This is generally positive, but can be detrimental if developers find questionable requirements and are reluctant to report them. That is one reason that it is desirable to have other participants, such as software security engineers, in advisory roles throughout this phase. Such participants also need to have enough influence to effect their demands. While there might be contention between developers and these other groups attempting to have their requirements included, the result will likely be an increase in the overall value of the system to the various group members.

Verification

The goal of those verifying the system is to ensure that it is being constructed according to detailed customer and user specifications both in terms of accurately depicting those requirements and ensuring that all requirements have been included. This means that individuals not directly involved in creating the programs must go through each and every requirement and be shown how each one has been implemented. They also must determine that the proposed methods of testing the programs against requirements and specifications meet the needs. The tests themselves are done in the validation phase.

Part of this phase should also be the determination that the set of requirements are complete and correct. If not, there needs to be a process whereby the results of the verification are fed back to the appropriate groups involved in the requirements phase. If any requirements are changed as a result of this review, then specifications must be updated and changes made to the program to incorporate the corrected set of requirements. It can easily be seen how change emanating from later phases in the life cycle are so much more costly in money and time to implement, and how it is so much less burdensome to get it right early in the life cycle.

Thus developers are participants, but not the decision-makers with respect to verification, as shown in Table 11.2. In some sense, the reviewers are encouraged to find errors and omissions and to ensure that the errors are corrected. Again, there is some contention involved because of having to rework the programs. This can delay completion and increase resource use. However, overall there is enormous benefit from the verification phase since errors and omissions caught at this phase are orders of magnitude easier and less expensive to fix than if the errors are caught later in the life cycle. Any costs or delays caused by verification should be set against the much greater costs and delays that would be incurred as a result of missing the errors.

Validation

In this phase, the correct working of the software system is proven by running a series of prespecified tests and observing the results. If errors are detected, they are fed back to the developers for reworking of the computer programs. In turn the revised programs need to be retested until the errors are resolved.

At issue here is whether so-called security and safety functional testing, in which the testers check that the software system is not doing that which it is not supposed to, and nonfunctional testing, in which security services, safety, performance, reliability, resiliency, and the like, are sufficiently represented in the testing program. Often they are omitted, which is why it is important to bring in those with vested interests in these other characteristics, as shown in the validation phase in Table 11.2.

Deployment, Operations, Maintenance, and Technical Support

There is quite a range of ways in which software can be deployed ranging from downloading from a Web site to custom installation. The former method usually involves some form of online or telephone support, typically from a call center or help desk, with more difficult situations being referred to technical support. Depending on the nature, size, cost, and complexity of the software system, the vendor might have their field service organization manage the installation, or the customer's technical support engineers might have the ability to do the installation in the field, particularly if it is an upgrade rather than a new installation. Ongoing operations will usually be managed by the organization running the software system using their internal staff or a third party, and scheduled maintenance and repair services will generally be done by the vendor's field service engineers or a third-party original equipment manufacturer (OEM).

Generally, information security and safety software engineers, either from the vendor or customer organization, will not be involved in the operational phase unless there is a security breach or a safety-related incident. However, there are strong indications that there should be security and safety oversight of the maintenance and repair functions since, for example, there are a number of recorded cases of counterfeit parts being substituted, some of which contain malware. In Table 11.2, we show these participant requirements along with others that might be called in to perform forensics and track down and arrest the perpetrators in the event of a breach or the discovery of malicious software or hardware.

The operations phase is probably the one that receives the least relative attention from security and safety proponents. The operations and technical support staff generally do not have the expertise to deal with security incidents, yet the vast majority of security-related attacks are on software systems that are in production. For safety-critical control systems, operational staffs are generally trained in how to handle day-to-day hardware malfunctions and failures, but are generally much less well equipped to detect and deal with software and data breaches, as the Stuxnet attack in Iranian uranium-processing facilities demonstrated.

Decommissioning and Disposal

The area that generally receives the least attention in the private sector is the decommissioning and disposal of security-critical and, to a lesser extent, safety-critical software systems and the data that they might contain. Very few texts on software security even discuss the disposal phase and the consequent risk of leakage of intellectual property and other sensitive data. On the other hand, government agencies in particular, such as the U.S. Department of Defense, often address the disposal of safety-critical systems—especially if the decommissioned system contains dangerous components that could kill or injure—and classified security-critical data.

If the responsibilities for decommissioning and disposal are clearly defined and assigned, those doing the job have some incentive to perform correctly and completely, since failure to perform might be readily discerned were an incident with, or data leakage from, a discarded system to occur. The greatest risk here, especially for security-critical systems, is that there may well not be anyone held accountable for the proper disposal of systems and media, so that there are few, if any controls in place. It is therefore important that security and safety experts, as well as internal and external auditors, be involved in the decommissioning and disposal phase, as shown in Table 11.2.

Overall Impression

The above analysis serves to point out that the exclusion or delayed participation of those who have specific interests, particularly in nonfunctional areas, leads to software system deficiencies in the early stages that result in subsequent excessive costs to revise requirements and solutions at later stages—the bolt-on versus built-in syndrome. The challenge is a matter of inducing those in authority to include all the necessary players early enough and sufficiently intensely to ensure that the overall value of the software system is maximized. Clearly it is not adequate to appeal to those in authority to do "the right thing." There have to be mechanisms for encouraging behavior that will push the system towards higher value. This can be done by mandate, threats, cajoling, economic plays, laws, regulations, and so forth. However, unless the impact is measured against the personal values of those who can influence the situation, it is not clear how effective these measures are. In the following section, we look at motivators and deterrents and what their impact might be.

Methods for Encouraging Optimal Behavior

There are a number of ways in which the behavior of individuals with respect to their roles and responsibilities throughout the SSDLC can be influenced in order to encourage the inclusion of certain attributes, such as security and safety. Such motivators and deterrents act upon individual decision-makers and project participants in order to encourage them to respond in ways that are globally desirable to the software-system manufacturing organization and other stakeholders, such as customers, business partners, regulators, and auditors.

Of course, decisions relating to the global optimization of the critical attributes are also highly subjective, but for our purposes, we will assume that it is possible to define some global value function that includes tangible and intangible costs and benefits from the perspective of an overall authority. While far from ideal, this approach at least enables one to determine which attributes are key, what their optimal levels might be, the contribution to global value of each phase of the SSDLC, and the specific participants and decision-makers who need to be persuaded, either directly or indirectly, to act in ways that promote the common good.

Pricing

If it can be applied, the pricing of resources can be one of the more efficient ways in which one can influence behavior. The typical mechanism for most organizations and their projects is some form of budgeting combined with chargeback. A critical component of this approach is to be able to fund certain desired attributes from a separate pool of money, the use of which does not impinge upon a particular project budget. This is important because, as described above, including security considerations and capabilities in the development life cycle is often against the personal interests of designers, developers, and testers who are generally motivated to complete a project on time and within budget. Security attributes usually fall into this category and will generally be underserved unless they are "paid for" from an organization-level.

Chargeback

The rules for charging overhead costs such as general administrative expenses can greatly influence the cost of a project and sometimes whether a project will actually go forward. Chargeback also influences the behavior of users. The chargeback rules affect how participants respond to various requirements. It is often desirable to have a central pool of funds for such factors as information security. In this way, senior management can enforce security requirements, say, without penalizing the development team for any delays or cost overruns due to incorporating security attributes into the software system.

Costs and Risk Mitigation

One can invoke incentives, such as the linking of a person's compensation to how well they implement corporate goals in areas such as security, safety, and resiliency. The latter goals might not normally be success factors for participants such as developers and testers.

The behavior of some can be influenced by opportunity costs and costs of opportunity. The former, which are those costs incurred in doing one thing rather than another, imply that a decision-maker might be in a position to choose one means of risk mitigation over another. An example would be the incorporating of security during the SSDLC or installing application firewalls and other tools once the software system has been deployed. In another example, a company might choose to invest in ruggedizing a data center by providing onsite backup for their systems, redundant equipment, emergency power supplies (e.g., battery backup, diesel generators), several telecommunications vendors, and so on, rather than spend the money on a fully functional backup data center. With some of the money saved, the company might contract with a third party to provide disaster recovery services on demand. The latter influence—namely, the cost of opportunity—refers to the willingness of some to make decisions that reduce current costs but which might have an adverse impact at some future time. This is termed risk acceptance, where for example the decision-maker is willing to take a chance as to whether a cyber attack will happen.

Management Mandate

An effective means of encouraging the desired behavior is to make it company policy and put in place credible enforcement procedures. If senior management is convincing and has a track record indicating that they mean what they say, then this approach can be extremely effective. A very good example of this is the previously mentioned case of Bill Gates, founder of Microsoft Corporation, who made a pronouncement that trustworthiness would be a key ingredient in their products going forward. In the decade since that policy statement, Microsoft has greatly increased its reputation for building secure software.

By establishing a goal and publishing a related policy, senior management is directing lower-level managers to meet or beat the goal but is leaving the details of implementation to the judgment of the managers. Usually this means that reasonable budget requests for introducing security, safety, and other capabilities will be met with approval.

Legislation

Lawmakers have become highly sensitized to the threat of data breaches leading to identity theft because of the alarming number of breaches reported and the fact that many of their constituents have been victims of identity theft and account takeover. In the United States, most of the states have enacted data privacy laws but as of this writing, several Federal bills have been crafted in the U.S. Congress but none have actually passed into law.

Laws can be a very effective means of motivation or deterrence in many slower-moving areas. However, the challenges of cybersecurity and system safety are so dynamic that it is difficult for even tech-savvy individuals to keep up with such rapid change, let alone members of Congress, many of whom do not have a good understanding of information technology and especially the security-related and safety-related consequences

Regulation

Regulators translate laws into actionable directives and then ensure that the resulting regulations are being followed through various oversight, monitoring, and auditing methods. In the U.S. financial services industry, banks are regulated by a number of agencies, including the Federal Reserve Board and the Office of the Comptroller of the Currency (OCC), and securities firms and all public companies in the United States are regulated by the Securities and Exchange Commission. Some agencies have issued very detailed guidance in regard to application security, such as OCC Bulletin 2008-16, which includes many suggestions as to how to achieve acceptable levels of application security [6].

Many other industries, of which some are involved with developing and using both security-critical and safety-critical software systems, are regulated, such as the telecommunication industry that is regulated by the Federal Telecommunications Commission and the airline industry by the Federal Aviation Administration. These regulators often provide standards that must be met by aircraft manufacturers, ground vehicle manufacturers, nuclear power plant builders, and so on.

Standards and Certifications

Not only must many industry members comply with laws and regulations, they might need to achieve a predefined certification level in order to operate within that industry. Certifications are common when it comes to safety-critical software systems, which are usually in the form of industrial control systems that could cause loss of life or injury if they malfunction or fail. Obtaining certifications are a very strong motivator.

While common in industries operating safety-critical systems, certifications are few and far between when it comes to security-critical systems. The Payment Card Industry-Data Security Standard (PCI-DSS) provides application security guidance, but its jurisdiction is limited to issuers and processors of payment card information, such as data relating to credit, debit, and cash cards [7].

It is to the detriment of many security-critical software systems that standards and certifications, equivalent to those that apply to safety-critical systems, are not available to those developing, operating, and maintaining such security-critical systems.

Going Forward

In order to deal with the shortfall in safety and security in software systems, it is necessary not only to implement the recommendations contained in this book but also to look to promising developments, some of which are in the research phase and others that have been implemented commercially. This is not to endorse any of these approaches but to claim that, in addition to the synergies of transferring good practices between the software safety and security engineering disciplines, we need to look beyond what is currently being done as there is clearly room for improvement given the many breaches, malfunctions, and failures that occur with regularity.

Confirmation that there is a need for breakthrough approaches and technologies is readily apparent from the previously mentioned Broad Agency Announcement (BAA) [8] issued in February 2012 by Defense Advanced Research Projects Agency (DARPA) with the title "High-Assurance Cyber Military Systems (HACMS)." It is noteworthy from the quotation below, which was taken directly from the BAA, that the definition of high-assurance cyber-physical systems includes requirements to satisfy specifically "... safety and security properties" in addition to their being "functionally correct."

> The goal of the High-Assurance Cyber Military Systems (HACMS) program is to create technology for the construction of high-assurance, cyber-physical systems, where high assurance is defined to mean functionally correct and satisfying appropriate safety and security properties. Achieving this goal requires a fundamentally different approach from what the software community has taken to date. Consequently, HACMS will adopt a clean-slate, formal methods-based approach that enables semi-automated code synthesis from executable, formal specifications. In addition to generating code, such a synthesizer will produce a machine-checkable proof that the generated code satisfies the functional specification as well as security and safety policies.

Perhaps the most telling assertion in the introduction section of the report is that in order to construct high-assurance cyber-physical systems, there is a need to develop and apply a different approach from what has existed until now. DARPA insists that "... research should investigate innovative approaches that enable revolutionary advances in science or systems."

While there is clearly a need for innovation in this area, it is not obvious as to what forms such development technologies should take. It is worth the effort to try to follow the particular research that will emanate from this BAA and that will be funded by DARPA in order to get some guidance as to future directions in creating safe and secure systems.

However, in the meantime, there are a number of interesting approaches to improving the safety and security of software systems, two of which will be discussed here. These are relatively new methods and processes but hold promise for improving the safety and security of critical systems and systems of systems. One relates to tampering; the other to so-called "patterns." The former is in the category of avoiding the impact of attacks by shielding (or immunizing) software systems from attacks, whereas the second is a process for information sharing and learning from experience, which has already been applied in the security areas and holds promise with respect to safety attributes.

Tampering

It is clear from the frequency and descriptions of attacks, both successful and not, on software-intensive systems, and from the manner in which they are discovered, that much of the time victims don't even know about most cyber attacks and are also ignorant about quite a few physical attacks. It is not always clear whether victims' lack of knowledge about attacks having taken place stems from not having detected them or because the criminals are expert enough to cover up their tracks—or both.

Experience in information security and attempts to prevent loss of valuable data—commonly called data leak prevention (DLP), makes one aware of the differences among tamper-evident, tamper-resistant, and tamperproof methods and how they might apply to security-critical and safety-critical software systems and the data they contain and handle.

Tamper Evidence

In simple terms, tamper-evident means that attempts to gain access to software or hardware systems or media is apparent from some change in the protective logical or physical "tape," which the thief is not able to reverse to remove evidence of tampering. Often, the individuals doing the tampering do not particularly care whether the recipient of a package or the software knows that someone has broken in. The recipient may have some claim on the sender or issuer of the product, but that is about the limit.

In other cases, the owner of a package or a piece of luggage, say, may have to allow some known authority, such as the Transportation Security Administration (TSA), but only that authority, to access the contents. Special padlocks have been designed to permit access by TSA representatives and show whether the package has been opened by someone with a TSA standard key.

Tamper Resistance

Tamper resistance is a different concept. Here access to the product may be difficult but not impossible. In the physical world, those funny little screw heads, come to mind. They allow someone to turn the screw in a clockwise direction with a regular screwdriver but won't allow counterclockwise unscrewing come to mind. There are quite a few approaches in the physical world. Comparable systems for software might also exist, although if they do, they are not well-known. In any event, with some effort that is likely to be destructive, such as drilling out the screws, the protective method can be bypassed. This is good for someone interested in getting at the innards of the device and not worried about whether such intrusion can be detected by intended users.

Tamper resistance works well to deter or discourage the amateur thief but can be broken by professionals with the appropriate skills and tools. Depending on the value of the assets being protected, such tamper resistance might well be adequate, but many not meet requirements for protecting highly sensitive data or classified equipment.

Tamperproofing

Tamperproofing is different from the other methods. Here, attackers are usually not only hampered in their attempts to gain access to a working version of the hardware or software being protected but the hardware or software may self-destruct. There are two flavors of tamperproofing. In the physical world in particular, attempts to break into the item will result in the immediate destruction of whatever is contained within the tamperproof container. The destruction might be through shorting and burning out of circuitry and wiping out of any software and firmware contained therein, or the destruction can be more aggressive, and the unit can be made to explode—but in such a situation the risk of collateral damage and injury to bystanders needs to be considered.

From a software perspective, there are methods and products that can prevent anyone from accessing and modifying software. Since it is readily apparent that few, if any, software systems can be fully protected from attack and that deterrence is seldom effective, we're left with avoidance. Avoidance is not about making the assets available to a potential attacker. This is acceptable in some situations, but not all. Obviously, if you need to run a program or process certain sensitive data, then one has no choice but to make the program and data available, knowing that attack is virtually impossible to prevent. Much as in the physical world, the answer lies in building up the immune system rather than trying to destroy the attacks and attackers. Tamperproofing is a very effective way to do so. Of course, a decision has to be made as to whether the self-destruction of specific software and hardware is justifiable. The decision for software is usually much simpler in that it is easy to restore copies of software. There may be a reluctance to destroy extremely valuable equipment and a preference for disabling and presevering it so that it can be brought back to operational status at some future time. For example, one would not want such systems as those involving warfare (artillery, planes, ships, submarines, etc.) to fall into enemy hands, so self-destruction is a viable approach. In other cases, it might be adequate to get assurance that someone who has captured such equipment is not able to use it or reverse-engineer it. The latter aspects are very difficult to prove.

In Table 11.3 we show advantages and disadvantages for the different levels of tamper-related characteristics. Essentially, the cheaper, easier-to-implement methods, such as those that provide evidence of tampering, offer little protection of the contents but may act as a deterrence since the tape protecting the box, say, will show that someone tried to or succeeded in getting unauthorized access to the contents. Furthermore, it behooves the recipient to check that the

Table 11.3
Advantages and Disadvantages of Various Tamper-Related Methods

Characteristic	Advantages	Disadvantages
Tamper-evident	Inexpensive and easy to implement Provides evidence of attempted and/ or successful break-in	Easy to access system, but may be subject to penalties, etc., such as invalidating warranty Does not prevent intrusion and access Assets become available in usable form
Tamper-resistant	Usually inexpensive and easy to implement Usually provides evidence of tampering Provides some measure of protection, much like a physical roadway speed bump	Moderately easy to access system if some level of destruction is tolerable by tamperers Does not prevent intrusion and access Assets become available in usable form
Tamperproof	Ensures that attacker does not get access to working versions of hardware or software in usable form Intellectual property and secrets may be assured if self-destruction is effective Usually provides evidence of tampering	Protected hardware is destroyed so that it cannot be reused by originator Need tailored software that might be costly and difficult to create and implement

contents are complete and intact when there is an indication that someone has opened the container. More protective methods are generally more expensive and difficult to implement but offer a higher level of assurance that the sensitive and valuable resources will be safe from unauthorized access. The self-destruction mode is the most effective, but makes units useless, as intended, so that if a unit is recovered it has no reclamation value to the rightful owner either.

In Table 11.4 we show which tamper-related methods are available for security-critical and safety critical software, hardware, and media.

A Brief Note on Patterns

The predominant issue, which comes from examining what is needed to build software systems that are both safe and secure, is the need to communicate among the various stakeholders and interested parties across silos in order to have the desired attributes and characteristics added to systems during the early stages of the SSDLC. While it might be helpful at some level to pronounce a normative approach, such as including all relevant stakeholders in meetings to discuss all the required safety and security attributes, that really does not address the question of which mechanisms are available and should be used to facilitate the transfer of information.

In an attempt to address this issue researchers have developed the concept of "patterns" that can be used to convey to designers and developers the experience of those who have addressed the same particular issues in the past. According to Schumacher [9], the interest in the use of patterns was created by Gamma [10] in 1995. The Schumacher book specifically addresses "security patterns," which are highly relevant to this book, and it is easy to see how this approach can be readily extended to cover other nonfunctional features such as safety, resilience, and the like.

According to Schumacher, patterns have the following properties:

- A pattern describes both a process and a thing—where the "thing" is a high-level design outline or detail of programming code, and the "process" is a set of instructions for creating the defined configuration effectively.

- A pattern is a solution that is both proven and of high quality and resolves a given problem optimally.

- A pattern supports both the understanding of a problem and its solution.

- A pattern is generic—it does not describe a specific solution or arrangement of components.

Table 11.4
Tamper-Related Methods by Security-Critical and Safety-Critical Software Systems

Characteristic	Security-Critical Systems		Safety-Critical Systems	
	Software	**Hardware/Media**	**Software**	**Hardware/Media**
Tamper-evident	Monitor/review attempted logons, for unauthorized users in particular Instrumentation in programs to monitor and report unusual use of software Instrumentation for detecting unusual data access patterns	Tamper-evident tape (e.g., over screws used to close equipment) Release of dye or other substance when security devices are broken Tamper-evident tape on cartons of tapes, paper, etc., and on individual tapes, etc.	For physically isolated systems, report violations of policy and procedures for physical access to end-user devices, servers, etc. For logically isolated systems, report attempts to break into software and stored data, and networks and transmitted data	Monitor attempts made to make control system malfunction or fail via physical or logical means
Tamper-resistant	Implementation of strong authentication technologies to access applications and data Management of authorizations to keep them current and accurate—effective change control Defense in depth using firewalls, intrusion detection, etc.	Use one-way screws or fasteners needing special tools in order to restrict ability to open equipment and introduce malware and otherwise pervert or destroy the software Physical locks on media devices	Use strong methods for authentication and data encryption Require physical presence (i.e., no remote access and no use of USB and similar devices) Defense in depth using firewalls, intrusion detection, etc.	Use one-way screws or fasteners needing special tools in order to restrict ability to open equipment and compromise existing circuitry or insert rogue circuitry
Tamperproof	Strong encryption of software and data, particularly for mobile devices and media Self-erasure of software and data when unauthorized access attempts are detected	Shielding that burns out circuitry if broken Media that cause data to be erased in response to attempts made to read or change the data by unauthorized individuals	Physical isolation of systems and data (e.g., air gaps between external and internal networks)	Shielding that burns out circuitry if broken Media that cause data to be erased in response to attempts made to read or change the data by unauthorized individuals

- A pattern tells "a successful engineering story" and initiates a dialog about how to best resolve a particular problem.
- Patterns frequently tackle problems in indirect, unusual, and counter-intuitive ways.

The concept of patterns and their use is somewhat analogous to the use of objects in object-oriented programming in that concepts are encapsulated in reusable form so that software system engineers do not have to start from the beginning each time that they need to address a particular requirement. In this way, they can benefit from the experience of their predecessors who addressed the same problem previously and came up with an effective solution.

Such an approach could potentially provide a very effective mechanism for software security engineers and software safety engineers to communicate with one another and with other members of the various SSDLC teams. And, as discussed throughout this book, it is communication among subject-matter experts that is a critical success factor in building and operating safe and secure software systems.

Conclusions

Having journeyed through the evolution and requirements for security-critical and safety-critical software systems, one is impressed with the tremendous amount of good material that is already available in this space, whether it is in the form research papers, commercial tools, scholarly and practically oriented books, and government and industry standards and certifications. Yet, for a variety of organizational, cultural, and technical reasons, the strengths in various aspects and zones of software systems engineering never seem to find their way to other areas. There will be considerable potential benefits to the high-assurance software world through just the transference of existing knowledge about software security and safety processes, methods, tools, and the like across the entire field. Today's gaps between software security and safety engineering are exacerbated by the fissures between government and the private sector, and the private sector and academia and research institutions. Bridging those gaps will be no easy task. Prior attempts at public-private information-sharing and collaboration have met with limited success, and there is no indication that the present situation will improve in the foreseeable future unless effective legislation and regulation were to come into being, which itself is not particularly likely. There is still a need to express the risks of attack, malfunction, and system failure in a way that is understood by lawmakers, regulators, law enforcement, the military, and the public. Also these groups need to internalize and personalize those risks and understand what it will take to mitigate them. Such education and awareness is critical to the formulation, implementation, and enforcement of effective laws, regulations, policies, and certification requirements.

Yet, even if everyone "spoke the same language" and struck an appropriate balance between protection and protectiveness, between the relatively casual testing of information systems compared to the rigorous verification and validation of safety-critical control systems, and between the strict certification standards in avionics, for example, and the lack of standards for most business software-intensive systems, there would still be big holes in the approaches. Not enough effort has been applied to date or appears likely to be made in the near future. The research, development, and implementation effort, which has been and is being made, appears to many to be grossly inadequate as witnessed by the large and escalating numbers of successful attacks and the many complaints by those experts who should know about the deficiencies in the software system development life cycle processes and content.

This latter dangerous situation draws others to the conclusion that the only way to succeed in this arena will be through breakthrough innovation. That may indeed be what is needed, but there is no guarantee that currently proposed and existing research projects, whether in the public or private sectors, will yield results that will be strong enough, cost-effective, and timely enough to get ahead of the attackers. Meanwhile, there are today's security and safety issues to address, and they cannot wait for some hoped-for white knight approach to save us from what appear to be some inevitable disasters.

This conclusion brings us back to the need to form a solid body of knowledge across the software safety and security engineering disciplines and to implement the most effective methods and procedures across the entire software security and safety space. In addition, effective enforcement, measurement, management, and response capabilities need to be developed—they alone will lead to significant improvements, but are by no means the whole answer.

In a nutshell, the goal, with respect to engineering safe and secure software systems, should be to learn from the past, bring together team members with the appropriate state-of-the-art technical and business knowledge, and push for innovative and practical research into approaches and solutions for the future. The problems surrounding the making and operating of safe and secure software-intensive systems will only increase over time. Therefore it is necessary to develop the means of resolving these problems at an accelerating rate if there is to be any hope of catching up with and eventually surpassing the rate of increase of threats, exploits, and vulnerabilities. We are talking here of a veritable Manhattan Project for the engineering of safe and secure software systems [11].

How quickly that crossover point, where the defense against cyber attacks and the malfunctioning or failure of safety-critical software systems overtakes the dynamics of attacks and deficiencies, will be reached depends on the level of effort and resources committed to the task. It is unreasonable to expect that this point at which attackers and defenders are on equal footing will occur in anything less than 5 years, assuming major commitments by those who understand the risks and have the funds, power, and authority to make it happen. If such support is not forthcoming, then defenders may never catch up and we will be forever condemned to suffering increasing damaging attacks, breaches, and failures. And if there is support, but it is not sufficient to overtake the abilities of attackers and their attacks as well as unintentional and uncontrollable events, then there is the risk that catastrophic events, man-made or natural, will occur before adequate security and safety measures are in place, which could in effect negate the entire effort. It is never too soon to get high-level commitments to go forward with the necessary efforts, but might it already be too late?

Endnotes

[1]   Allman, E, "Managing Technical Debt," *Communications of the ACM,*" Vol. 5, No. 5, May 2012, pp. 50–55.

[2]   See B. Gardiner, "We're All Climate-Change Idiots," *The New York Times,* July 21, 2012.

[3]   See http://commerce.senate.gov/public/index.cfm?p=Hearings&ContentRecord_id= 676548f-a2a7-40ff-a18d-889a7907801c#hearingParticipants, accessed on July 23, 2012.

[4]   See http://latimesblogs.latimes.com/world_now/2012/06/japan-tsunami-nuclear-disaster-fault-apology-blame.html, accessed on July 23, 2012.

[5]   Axelrod, C.W. "A Dynamic Cyber Security Economic Model: Incorporating Value Functions for All Involved Parties," in *Threats, Countermeasures, and Advances in Applied Information Security*, M. Gupta, J. Walp, and R. Sharman (eds.), Hershey, PA: IGI Global, 2012.

[6]   Available at http://www.occ.gov/news-issuances/bulletins/2008/bulletin-2008-16.html, accessed on July 23, 2012.

[7]   Available at https://www.pcisecuritystandards.org/security_standards/documents.php?document= pci_dss_v2-0#pci_dss_v2-0, accessed on July 23. 2012.

[8]   See https://www.fbo.gov/index?s=opportunity&mode=form&id=925ab03b1bb59b1ac48 4e5a240b77097&tab=core&_cview=0, accessed on July 23, 2012.

[9]   Schumacher, M., et al., *Security Patterns: Integrating Security and Systems Engineering*, Chichester, England: John Wiley, 2006.

[10]   Gamma, E., et al., *Design Patterns—Elements of Reusable Object-Oriented Software,* Boston: Addison-Wesley Professional, 1995.

[11]   As is commonly known, the Manhattan Project was a huge effort to develop nuclear bombs during World War II, http://en.wikipedia.org/wiki/Manhattan_Project, accessed on July 23, 2012.

Appendix A
Software Vulnerabilities, Errors, and Attacks

In this appendix, we discuss commonly available lists of attacks, vulnerabilities, and risks relating to software, predominantly web applications. Web applications generally have a strong security bias, as they are information systems rather than control systems and, being web-facing, they are prone to attacks by external agents. However, we will see that the lists contain recommendations that are also applicable to information systems that do not face the web and to safety-critical control systems, even though they were not necessarily created with such systems in mind.

As shown in Figure A.1, there are two main approaches to making developers and testers, in particular, aware of the security and safety risks relating to software systems engineering. One is to bring top vulnerabilities and risks to their attention. These lists are best represented by the popular OWASP [1] top ten vulnerabilities and security risks, shown on the left side of Figure A.1, and by the CWE/SANS [2] top 25 most dangerous programming errors, also on the left of the diagram. Furthermore, we show a fair amount of overlap in the diagram between the two lists, about which we will be more specific later.

It is interesting to note that lists equivalent to the OWASP and CWE/SANS lists do not appear to exist for software safety. A list claiming to be "The Top Ten Internet Safety Tips from Net Nanny is really a set of security suggestions and recommendations on the use and blocking of various resources available on the Internet [3]. This is not a matter of the web applications being unsafe, but of their potential danger to misuse by children, which could lead to harm.

There are some safety-related rankings, such as the "Top Ten Safety Tips," from *Emergency Management Magazine* [4]. However, these tend to be high-level guidelines rather than specific technical suggestions, such as contained in the OWASP and CWE/SANS security-related rankings mentioned above.

On the right side of Figure A.1 we show some approaches to the enumeration and classification of weaknesses, vulnerabilities, attack patterns and malware attributes. These approaches tend to list all existing attributes rather than trying to rank them in order of impact. For example, the CWE/SANS Top 25 are taken from more than 900 CWE cases.

Ranking Errors, Vulnerabilities, and Risks

There are two schools of thought when it comes to publicizing the riskiest violations and the most dangerous programming errors relating to information security. One group claims that, by knowing where the greatest vulnerabilities lie, software-system designers and architects will account for them in their designs, and developers will avoid writing program code that will expose their computer applications to these weaknesses. In addition, testers and auditors have checklists against which they can use to ascertain that the programs do not violate the most egregious risk attributes.

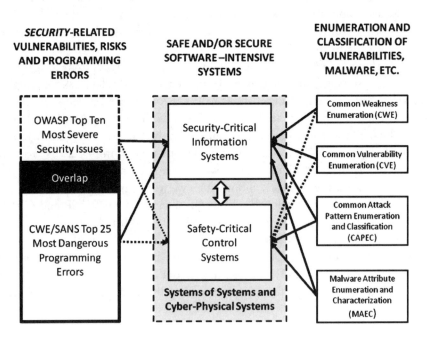

Figure A.1   Various approaches to addressing weaknesses, vulnerabilities and risks.

The other viewpoint is that developers will *only* attend to those vulnerabilities and errors on the lists, and will not attend to other important errors that did not quite make it to the list of top items; that is to say, the lists should certainly be viewed as *necessary* to follow, but they are not *sufficient* for ensuring a high-level of security or safety. Who is to say that the eleventh and twelfth ranking vulnerabilities are not extremely important even though they didn't make the top 10 or 25? To their credit, the CWE/SANS publication (wherein their lists are described in detail) guides readers to an "On the Cusp" page, which gives an additional sixteen weaknesses that did not fall into the top 25. On the other hand OWASP continues to limit its list to 10 items, although they do warn developers not to limit themselves to the 10 risks listed but to read various OWASP online documents to learn about the hundreds of issues that affect overall security.

The OWASP Top Security Risks

OWASP began publishing its "top 10" lists in 2003. In the early versions, the list consisted of attacks, vulnerabilities, and countermeasures. In 2007, the focus was entirely on vulnerabilities, and in 2010, the list contained security risks. Figure A.2 shows how the 10 examples in the list changed from 2003 to 2010. Five items, representing half of the total, were dropped during the 2003-to-2010 period. For many of the remaining items, their rankings changed between one version and the next. Sometimes these changes were quite significant; for example, the category "broken authentication and session manag-ement" jumped from seventh position in 2007 to third position in 2010.

Judging from past trends, a new version of the OWASP Top Ten might be expected to appear in the 2013 timeframe, although such timing has not been announced by OWASP.

The OWASP Top 10 has received a great deal of attention from information security professionals, even to the extent that regulators and oversight entities have suggested that the OWASP Top 10 should be a requirement for compliance [5].

In Table A.1, we list the 2010 Top 10 web application security risks. In the table, the occurrence of the item is explained and the potential impact is given for each of the security risks. OWASP published a report with the title *OWASP Top Ten—2010: The Ten Most Critical Web Application Security Risks*, which is available online [6]. The report describes each of the top ten security risks in detail.

For each item, the OWASP report lists the following:

• Threat agents;

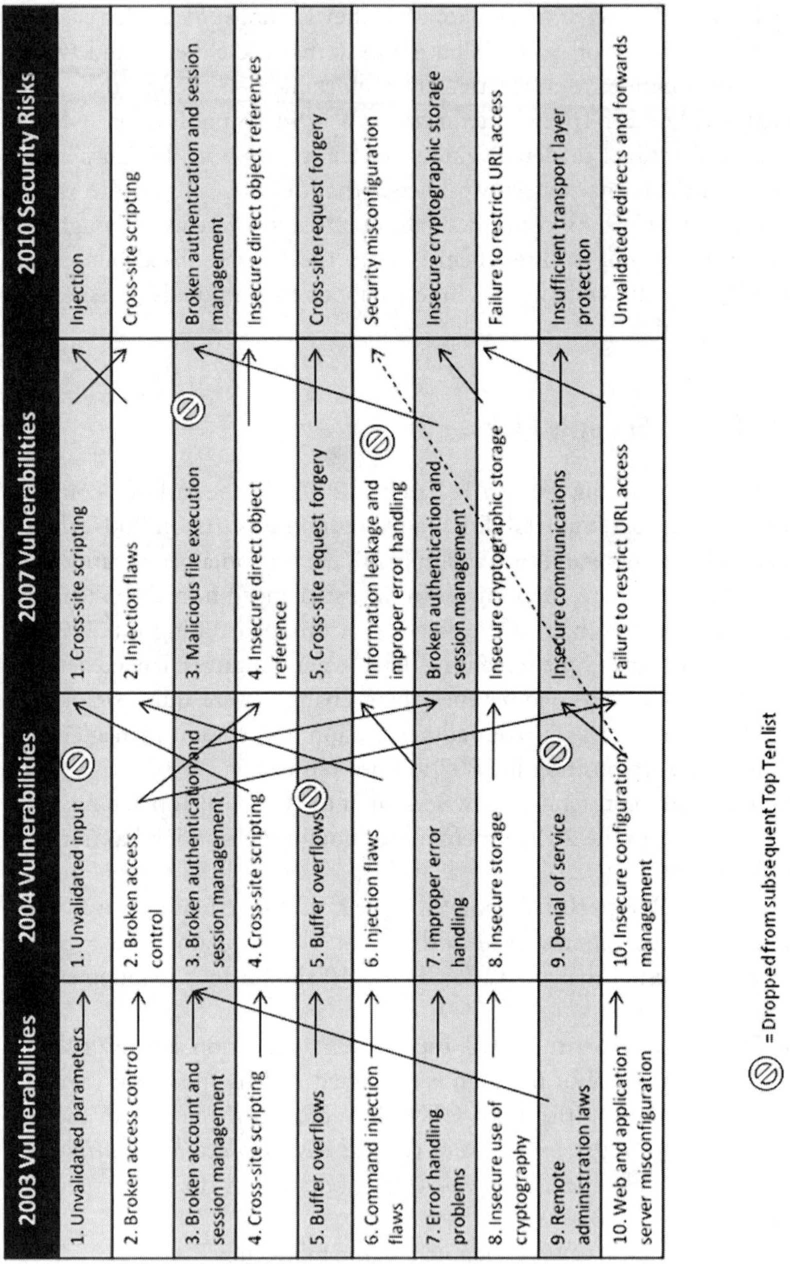

2003 Vulnerabilities	2004 Vulnerabilities	2007 Vulnerabilities	2010 Security Risks
1. Unvalidated parameters	1. Unvalidated input	1. Cross-site scripting	Injection
2. Broken access control	2. Broken access control	2. Injection flaws	Cross-site scripting
3. Broken account and session management	3. Broken authentication and session management	3. Malicious file execution	Broken authentication and session management
4. Cross-site scripting	4. Cross-site scripting	4. Insecure direct object reference	Insecure direct object references
5. Buffer overflows	5. Buffer overflows	5. Cross-site request forgery	Cross-site request forgery
6. Command injection flaws	6. Injection flaws	Information leakage and improper error handling	Security misconfiguration
7. Error handling problems	7. Improper error handling	Broken authentication and session management	Insecure cryptographic storage
8. Insecure use of cryptography	8. Insecure storage	Insecure cryptographic storage	Failure to restrict URL access
9. Remote administration laws	9. Denial of service	Insecure communications	Insufficient transport layer protection
10. Web and application server misconfiguration	10. Insecure configuration management	Failure to restrict URL access	Unvalidated redirects and forwards

= Dropped from subsequent Top Ten list

Figure A.2 Evolution of OWASP top ten from 2003 to 2010.

Table A.1

OWASP 2010 Top 10 Web Application Security Risks

Rank	Name	Occurs when ...	Potential Impact
1	Injection	... untrusted data is sent as part of a command or query.	Tricks the interpreter into executing unintended commands or accessing unauthorized data.
2	Cross-site scripting (XSS)	... an application takes untrusted data and sends them to a web browser without proper validation and escaping .	Allows attackers to hijack user sessions, deface websites, or redirect the web browser user to malicious sites.
3	Broken authentication and session management	... application functions related to authentication and session management are not implemented correctly.	Allows attackers to compromise passwords, keys, session tokens, or exploit other implementation flaws to assume other users' identities.
4	Insecure direct object references	... a developer exposes a reference to an internal implementation object, such as a file, directory, or database key.	Allows attacker to manipulate access control checks and other protection in order to gain unauthorized access to data.
5	Cross-site request forgery (CSRF)	... the browser of a logged-on victim is forced to send a forged HTTP request, which includes the victim's session cookie and other authentication information, to a vulnerable web application.	Allows attacker to force the victim's browser to generate requests that the vulnerable application believes are legitimate.
6	Security misconfiguration	... the configuration, which is defined and deployed for applications, frameworks, application servers, web servers, database servers, and platforms, is secure, with all settings defined, implemented and maintained, and all software and code libraries kept up-to-date.	Allows a broad variety of attacks to be successful.
7	Insecure cryptographic storage	... web applications do not protect sensitive data, such as social security numbers, credit-card information, authentication credentials, and intellectual property, with appropriate encryption or hashing.	Facilitates the theft and/or modification by attackers of weakly-protected data to conduct identity theft, fraud, industrial espionage, and other crimes.
8	Failure to restrict URL (uniform resource locator) access	... web applications do not perform access control checks each time pages are accessed.	Enables attackers to forge URLs to access hidden pages.
9	Insufficient transport layer protection	... web applications fail to authenticate, encrypt and protect the confidentiality and integrity of sensitive network traffic.	Allows attackers to "sniff" the data flowing over networks and obtain sensitive information such as usernames, passwords, social security numbers, and the like.
10	Unvalidated redirects and forwards	... web applications redirect or forward users to other pages and websites, using untrusted data to determine the destination pages.	Allows attackers to redirect users to phishing (sending messages from criminals masquerading as legitimate senders) or malware (malicious software) websites, or access unauthorized pages.

- Attack vectors and their exploitability;
- Security weaknesses and their prevalence and detectability;
- Technical and business impacts.

The OWASP report helps readers determine whether specific applications are vulnerable to each of the various threat agents, and how to prevent the threat from exposing the particular application to compromise. Examples of attack scenarios are also given. Merkow [7] provides very good explanations of the 2010 OWASP Top 10.

The CWE/SANS Most Dangerous Software Errors

The CWE effort is described as a "community-developed dictionary of software weakness types," [2]. The CWE effort is also in the category of approaches that enumerate exhaustive lists of weaknesses, vulnerabilities, threats, errors, and so on, and is included on the right side of Figure A.1. The top 25 most dangerous software errors for 2011 are available on the CWE website [8].

As of Version 2.2, there are 909 CWE errors. Included among these CWE errors are the OWASP Top 10, as well as the CWE/SANS Top Twenty-Five. By following the links provided to each of the OWASP Top Ten, one can see which CWE errors relate to each item. Those for the top 25 CWE errors are shown in the right-most column of Table A.2. Note that some of the CWE/SANS Top 25 do not have OWASP equivalents, and some have two.

Table A.2 shows the ranking not only of the highest-ranking 25 software errors but also of the next-ranked 16 errors. The CWE website [9] also splits the top 25 items into three categories, depicted in the table as columns A, B, and C, as follows:

- *Column A*: Insecure interaction between components (insecure way in which data are sent and received between components, modules, programs, processes, threads, or systems).

- *Column B*: Risky resource management (software does not manage properly the creation, use, transfer or destruction of important system resources).

- *Column C*: Porous defenses (defensive techniques that are often misused, abused or ignored).

Top-Ranking Safety Issues

There do not appear to be any safety equivalents to the lists of leading risks, threats, errors, and the like that exist for software security. This is probably due to there not being the same need for such lists in the safety arena because there are many highly-detailed safety standards against which software systems must be certified, many of which are described by Herrmann [10].

Table A.2
CWE/SANS Top 25 Most Dangerous Software Errors

Rank	Name	A	B	C	OWASP Top Ten
1	Improper neutralization of special elements in SQL command	X			1
2	Improper neutralization of special elements in OS command	X			1
3	Buffer copy without checking size of input		X		
4	Improper neutralization of input during web page generation (cross-site scripting—XSS)	X			2
5	Missing authentication for critical function			X	3
6	Missing authorization			X	4, 8
7	Use of hard-coded credentials			X	
8	Missing encryption of sensitive data			X	7, 9
9	Unrestricted upload of file with dangerous type	X			4
10	Reliance on untrusted inputs in a security decision			X	
11	Execution with unnecessary privileges			X	
12	Cross-site request forgery (CSRF)	X			5
13	Improper limitation of a pathname to a restricted directory (path traversal)		X		4
14	Download of code without integrity		X		
15	Incorrect authorization			X	4, 8
16	Inclusion of functionality from untrusted control sphere		X		4
17	Incorrect permission assignment for critical resource			X	
18	Use of potentially dangerous function		X		
19	Use of a broken or risky cryptographic algorithm			X	
20	Incorrect calculation of buffer size		X		
21	Improper restriction of excessive authentication attempts			X	3
22	URL redirection to untrusted site (open redirect)	X			10
23	Uncontrolled format string		X		
24	Integer overflow or wraparound		X		
25	Use of a one-way hash without a salt			X	7
26	Allocation of resources without limits				
27	Improper validation of array index				
28	Improper check for unusual or exceptional conditions				
29	Buffer access with incorrect length value				
30	Inappropriate encoding for output context				
31	Use of insufficiently random values				
32	Untrusted pointer dereference				
33	Concurrent execution using shared resource with improper synchronization (race condition)				
34	Improper cross-boundary removal of sensitive data				
35	Incorrect conversion between numeric types				
36	NULL pointer dereference				
37	Improper enforcement of behavioral workflow				
38	Missing release of resource after effective lifetime				
39	Information exposure through an error message				
40	Expired pointer dereference				
41	Missing initialization				

However, it is clear from the OWASP and CWE/SANS lists that, while many of the items clearly apply exclusively to web applications, others can be readily related to safety software. Examples of those risks and errors that can be applied to safety-intensive software systems include those relating to authentication, authorization, and encryption, for example. Strong safety-critical software systems will have such attributes as encryption and authentication even if they are only intended to operate in physical and logical isolation. Besides being a good practice in and of itself, this approach puts such safety-critical systems in good stead for when they might be incorporated into cyber-physical systems.

Examples of the guidelines that are relevant to safety software (but *not* safety-intensive software systems) can be found in the blog, *Emergency Management Magazine* [4]. According to the blog article, the top five mistakes made in the selection of "public safety software" are:

1. Poorly defined goals and operational outcomes;
2. Trying to solve the wrong problem;
3. Poor requirements and request-for-proposal documents;
4. Lack of coordination, cooperation, and communication;
5. Not getting peer reviews.

This list is interesting in that it essentially focuses on the need for high-quality risk analyses and requirements definitions, and highlights the importance of collaboration and communication. These are all factors that are emphasized throughout this book.

Enumeration and Classification

We now turn our attention to the various classification and evaluation methods, some of which are shown in Figure A.1, on the right side of the diagram. We will add several other related approaches.

Different from the "top so many" lists, these approaches (shown in Table A.3) are usually exhaustive lists of vulnerabilities and weaknesses for legitimate and malicious software. They are mostly published online by The MITRE Group in the form of dedicated websites for each enumeration category. In Table A.3, we also include the focus of each method and the purpose of each approach.

The difference between a vulnerability and a weakness is given in the definitions below, where a vulnerability is the occurrence of a weakness, much as an exploit or attack is the occurrence of a threat.

Table A.3
Various Enumeration, Scoring, and Classification Methods

Acronym	Full Name	Focus	Purposes
CAPEC	Common Attack Pattern Enumeration and Classification	Attacks	To describe attack patterns associated with high-level malware taxonomy, such as those dealing with network reconnaissance, propagation insertion and command and control. To ensure that the attacker's perspective with respect to implementing these behaviors is properly represented. See http://capec.mitre.org/
CC	Common Criteria	Software security	To facilitate the comparison of the results of independent evaluations of security properties of information-technology products and systems using a common set of requirements for security functions. See http://www.commoncriteriaportal.org/
CCE	Common Configuration Enumeration	Configuration	Facilitates fast and accurate correlation of configuration data across a number of information sources and tools using unique identifiers for system configuration issues. See http://cce.mitre.org/
CEE	Common Event Expression	Malware	To describe logged events associated with malware activity in order to determine the presence of malware. See http://cee.mitre,org
CME	Common Malware Enumeration	Malware	To provide single, common identifiers for new and prevalent virus threats in order to reduce confusion during malware incidents.*
CPE	Common Platform Enumeration	Platform	To allow for assessing threats that a malware instance poses to organizational computing resources (applications, operating systems, and hardware devices) based on the targeted platforms. See http://cpe.mitre.org/
CVE	Common Vulnerabilities and Exposures	Vulnerabilities	To enable data exchange between security products and provide a baseline index point for evaluating coverage of tools and services. See http://cve.mitre.org/
CVSS	Common Vulnerability Scoring System	Vulnerabilities	To assess the severity of computer system security vulnerabilities by establishing a measure of how much concern a vulnerability warrants, compared to other vulnerabilities, so that efforts can be prioritized. See http://www.first.org/cvss
CWE	Common Weaknesses Enumeration	Weaknesses	To enable, from the CWE's unified, measurable set of software weaknesses, more effective discussion, selection, and use of software security tools and services, which can find these weaknesses in source code and operational systems, and better understand and manage software weaknesses related to architecture and design. See http://cwe.mitre.org
MAEC	Malware Attribute Enumeration and Characterization	Malware	To close the gap in malware-oriented communication using a language for characterizing malware based on its behaviors, artifacts, and attack patterns. See http://maec.mitre.org/
OVAL	Open Vulnerability Assessment Language	Vulnerabilities	To standardize how to asses sand report upon the machine state of computer systems. To provide enterprises with accurate, consistent, and actionable information so that they might improve their security. See http://oval.mitre.org
WASC TC	Web Application Security Consortium Threat Classification	Attacks	To develop and promote industry standards for classifying the weaknesses and attacks that can lead to compromise of websites, data and/or users. See http://www.webappsec.org/

The http://cme.mitre.org website has been replaced with the http://maec.mitre.org website (last accessed July 29, 2012)
The obsolete CME website contains the statement: "In late 2006 the malware threat changed away from the pandemic, widespread threats CME was developed to address to more localized, targeted threats, which significantly reduced the need for common malware identifiers to mitigate user confusion in the general public. Therefore, all CME-related efforts transitioned into support to MITRE's Malware Attribute Enumeration and Characterization (MAEC™) effort."
The websites in the table were all accessed on July 29, 2012.

WASC Threat Classification

The Web Application Security Consortium (WASC) is described [11] as being "made up of an international group of experts, industry practitioners, and organizational representatives who produce open source and widely agreed-upon best-practice security standards for the World Wide Web."

The WASC attacks and weaknesses listed in Table A.4 are linked to excellent descriptions of each attack and weakness [12]. Also available via the WASC website is the report, *WASC Threat Classification Version 2.0* (last updated as of January 1, 2010). The report is very extensive and examines each of the threats, attacks and weaknesses, which the report defines as follows:

- *Threat:* a potential violation of security (per ISO 7498-2).

- *Attack:* a well-defined set of actions that, if successful, would result in either damage to an asset or undesirable operation.

- *Weakness:* a type of mistake in software that, under proper conditions, could contribute to the introduction of vulnerabilities within that software ... [the term] applies to mistakes regardless of whether they occur in implementation, design or other phases of the SDLC. (From the MITRE Group's CWE website at http://cwe.mitre.org).

- *Vulnerability:* an occurrence of a weakness (or multiple weaknesses) within software, in which the weaknesses can be used by a party to cause the software to modify or access unintended data, interrupt proper execution, or perform incorrect actions that were not specifically granted to the party who uses the weakness. (From the MITRE Group's CWE website at http://cwe.mitre.org).

It is readily apparent that many of the WASC attacks and weakness are the same as in the OWASP Top 10 and the CWE/SANS Top 25. This raises the question as to whether the WASC threats can replace the CWE and CAPEC lists. In the frequently asked questions section in the *WASC Threat Classification Report* [11, 12] the response to that question is as follows:

> ... The work done by ... MITRE ... is far more comprehensive than anything online. The TC (threat classification) serves as a usable document for the masses (developers, security professionals, quality assurance) whereas CWE/CAPEC is more focused for academia ...

This is frequently the case. Research organizations and academic institutions tend to publish long and complex reports on this subject, but there is still

Table A.4
WASC Attacks and Weaknesses

Attacks	Weaknesses
Abuse of functionality	Application misconfiguration
Brute force	Directory indexing
Buffer overflow	Improper file system permissions
Content spoofing	Improper input handling
Credential/session prediction	Improper output handling
Cross-site scripting	Information leakage
Cross-site request forgery	Insecure indexing
Denial of service	Insufficient anti-automation
Fingerprinting	Insufficient authentication
Format string	Insufficient authorization
HTTP response smuggling	Insufficient password recovery
HTTP response splitting	Insufficient process validation
HTTP request smuggling	Insufficient session expiration
HTTP request splitting	Insufficient transport layer protection
Integer overflows	Server misconfiguration
LDAP injection	
Mail command injection	
Null byte injection	
It is readily apparent from the above OS commanding	
Path traversal	
Predictable resource location	
Remote file inclusion	
Routing detour	
Session fixation	
SOAP array abuse	
SSI injection	
SQL injection	
URL redirector abuse	
XPath injection	
XML attribute blowup	
XML external entities	
XML entity expansion	
XML injection	
XQuery injection	

a need for the academic work to be translated into a form that is more useful for the practitioner. This is accomplished to some degree by publishing an abbreviated list of leading cases as OWASP and CWE/SANS have done.

Summary and Conclusions

It is apparent from the above descriptions that there is a divide between the security and safety approaches to making practitioners more aware of common risks, threats, attacks, weaknesses, and vulnerabilities. In the security domain, we see the use of lists of leading issues and exhaustive lists of weaknesses, vulnerabilities, attacks, and so on. And yet these lists are guidelines with which software security engineers may or may not comply. There is little compulsion and even less enforcement, even when regulators support such lists.

On the other hand, software safety engineers have to adhere to strict standards if they are to have their systems certified at a particular level. This obviates the real need for lists of leading issues, and so we do not see them. This is all well and good, except for the fact that many of the security-related lists contain very good advice for those developing safety-critical software systems, especially when such systems are combined with information systems to form cyber-physical systems.

It serves both communities well if those responsible for security-critical software systems and those creating safety-critical software systems are aware of the tools available to the other group.

Endnotes

[1]   OWASP stands for the Open Web Application Security Project, which was founded in 2001. OWASP is an "open community dedicated to enabling organizations to conceive, develop, acquire, operate, and maintain applications that can be trusted." The website is www.owasp.org.

[2]   CWE stands for "Common Weakness Enumeration" and can be accessed at http://cwe. mitre.org. The SANS Institute is the "SysAdmin, Audit, Network, Security Institute," which is a leading training organization for information security professionals. The SANS Institute website is at www.sans.org . Both websites were last accessed on July 29, 2012.

[3]   See http://www.netnanny.com/learn_center/safety_tips, last accessed on July 29, 2012.

[4]   See http://www.emergencymgmt.com/emergency-blogs/tips/Top-five-mistakes-in-software-selection-010912.html, last accessed on July 29, 2012.

[5]   CC Bulletin 2008-16 on Application Security, see http://www.occ.gov/news-issuances/bulletins/2008/bulletin-2008-16.html, last accessed on July 29, 2012.

[6]   Available via a link at https://www.owasp.org/index.php/Category:OWASP_Top_Ten_Project, last accessed on July 29, 2012.

[7]   Merkow, M. S., and L. Raghavan, *Secure and Resilient Software Development*, Boca Raton, FL: CRC Press, 2010.

[8]   See http://cwe.mitre.org/top25, last accessed on July 29, 2012.

[9]   See http://cwe.mitre.org/, last accessed on July 29, 2012.

[10]   Herrmann, D. S., *Software Safety and Reliability*, Los Alamitos, CA: IEEE Computer Society, 1999.

[11]   See http://www.webappsec.org/, last accessed on July 29, 2012.

[12]   See http://projects.webappsec.org/w/page/13246978/Threat%20Classification, last accessed on July 29, 2012.

Appendix B
Comparison of ISO/IEC 12207 and CMMI®-DEV Process Areas

As can be seen from the Table B.1, there are many similarities among the majority of process areas of both ISO/IEC 12207 and CMMI®-DEV, although some comparisons are closer matches than others. Also, there are specific differences in categories—for example, verification and validation are included in the CMMI®-DEV category engineering, whereas ISO/IEC 12207 shows them as belonging to the software support category.

The main difference between these two sources is that ISO/IEC 12207 includes processes for individual phases of the development life cycle, whereas CMMI®-DEV does not. It appears that the reason for this is that CMMI®-DEV has a Specific Process (SP 1.2) to establish life cycle model sescriptions, whereas ISO/IEC 12207 does not, but rather presents a life cycle model directly.

Table B.1
ISO/IEC 12207 vs. CMMI®-DEV Process Areas

ISO/IEC 12207 Life Cycle Process Groups	ISO/IEC 12207 Process Areas*	Equivalent	CMMI®-DEV Process Areas	CMMI®-DEV Process Modules
Agreement	Acquisition process	←→	Supplier agreement management	Project management
	Supply process	←→	Supplier agreement management	Project management
Organizational Project-Enabling	Life cycle model management process	←→	Organizational process definition	Process management
	Infrastructure management process	←→	Organizational process definition	Process management
	Project portfolio management process	←→	Organizational performance management	Process management
	Human resources management process	←→	Organizational training	Process management
	Quality management process	←→	Organizational process performance	Process management
Project	Project planning process	←→	Project planning	Project management
	Project assessment and control process	←→	Project monitoring and control	Project management
	Decision management process	←→	Decision analysis and resolution	Support
	Risk management process	←→	Risk management	Project management
	Configuration management process	←→	Configuration management	Support
	Information management process	←→	Quantitative project management	Project management
	Measurement process	←→	Measurement and analysis	Project management
Technical	Stakeholders requirements definition process	←→	Requirements development	Engineering
	System requirements analysis process	←→	Requirements development	Engineering
	System architectural design process	←→	Technical solution	Engineering
	Implementation process	←→	Technical solution	Engineering
	System integration process	←→	Product integration	Engineering
	System qualification testing process	←→	Validation	Engineering
	Software installation process			Engineering
	Software acceptance support process			

Table B.1 (continued)

ISO/IEC 12207 Life Cycle Process Groups	ISO/IEC 12207 Process Areas*	Equivalent	CMMI®-DEV Process Areas	CMMI®-DEV Process Modules
Technical (continued)	*Software operation process*			
	Software maintenance process			
	Software disposal process			
Software Implementation	Software implementation process	←→	Technical solution	Engineering
	Software requirements analysis process	←→	Requirements development	Engineering
	Software architectural design process	←→	Technical solution	Engineering
	Software detailed design process	←→	Technical solution	Engineering
	Software construction process	←→	Technical solution	Engineering
	Software integration process	←→	Product integration	Engineering
	Software qualification process	←→	Validation	Engineering
Software Support	Software documentation management process	←→	Project monitoring and control	Project management
	Software configuration management process	←→	Configuration management	Support
	Software quality assurance process	←→	Process & product quality assurance	Support
	Software verification process	←→	Verification	Engineering
	Software validation process	←→	Validation	Engineering
	Software review process	←→	Verification	Engineering
	Software audit process	←→	Process and product quality assurance	Support
	Software problem resolution process	←→	Causal analysis and resolution	Support
Software Reuse	*Domain engineering process*			
	Reuse asset management process			
	Reuse program management process			

*Unmatched process areas are shown in *italic* type.

Appendix C
Security-Related Tasks in the Secure SSDLC

In this appendix, we list task areas for each of eight phases of the secure software system development life cycle (Secure SSDLC) in Tables C.1 through C.8, respectively. In Tables C.9 through C.16, we identify those who should be responsible for each task within each of the eight phases, and to what degree or level each area within an organization should be involved. Table C.17 provides a summary of responsibility levels for all eight phases and an aggregated level for the entire lifecycle.

While these tables contain substantial detail, they should still be used for guidance purposes only. Organizations can have very different organizational structures and risk profiles. Therefore the emphasis on which task areas to focus is likely to vary significantly from one organization to the next. That being said, there are usually more commonalities than differences when building security into software systems. Consequently, it should be helpful for those involved in creating and supporting security-critical software systems to have at hand what is essentially a checklist. This checklist should prove even more valuable to those who are less involved with security and are more focused on safety-critical systems; such individuals are less likely to be familiar with what is commonly required to make critical software-intensive systems secure.

It should be noted that we do not include any roles for software safety engineers in the secure SSDLC. This is fairly typical for information systems that are not directly subject to safety hazards. However, going forward, it is likely that more information systems will have safety-critical components or will be interfacing with safety-critical systems. Particularly when security-critical information systems are combined with safety-critical control systems to form cyber-physical systems do the creators, maintainers, and operators of the information systems have to account for safety issues. Of course, a physical event, such as a power blackout or a telecommunications network failure, can drastically impact information systems, but usually from performance, resiliency, and availability perspectives and not so much with respect to security.

As will be argued in Appendix D, the reverse is not necessarily true. Those designing, developing, operating, and supporting safety-critical software-intensive control systems often need to consider some security aspects of their software systems, even if the systems themselves are designed to be physically and logically isolated. There is always the possibility of intentional insider attacks by authorized employees or contractors, or unintended insider assistance to an attacker, as in the case of the Stuxnet worm that attacked the industrial control systems operating the centrifuges in uranium-processing facilities in Iran.

Task Areas for SSDLC Phases

In Tables C.1 to C.8, we provide descriptions and suggest purposes for each of the tasks throughout the following eight phases of the Secure SSDLC:

1. Risk analysis;
2. Security requirements;
3. Security architecture and design;
4. Development (building/configuration/integration);
5. Functional and nonfunctional security testing;
6. Deployment;
7. Operations, maintenance and support;
8. Decommissioning and secure disposal.

Tables C.1 through C.8 below will give those tasked with ensuring that security is built into software systems a fuller understanding of what is involved within each step (or task area) and why individual steps are needed.

The initial phase comprises the analysis of risk and its management. As shown in Table C.1, it is first necessary to determine potential risk factors, such as threats and weaknesses that evolve into exploits and vulnerabilities that have some probability of allowing successful attacks. These latter attacks cause losses, which can be reduced or mitigated to some extent using prevention, avoidance and/or deterrence methods.

The results of the risk analyses not only point towards the need to mitigate the identified risks, but also provide input to the requirements phase, as

Table C.1
Security-Related Tasks for the Risk Analysis and Control Phase

Task Areas	Descriptions	Purposes
Threat determination	Consideration of the environment in which a software system will operate and creation of a list of all the categories of relevant threats that might impact the system, including physical and logical threats. A list of specific threats within each category may be obtained from internal subject-matter experts, external services, industry and professional associations, and from various publications. Threat categories might include malware (i.e., viruses, worms), denial of service, network sniffing, spoofing (masquerading as someone else), identity theft, account takeover, perpetration of fraud, and theft of intellectual property (espionage). Specific individual threats might be a potential type of malware, such as a key logger, which captures each keystroke occurring on the victim's computer and transmits the keystroke data to the attacker.	To come up with a list by category of threats that might, if turned into exploits or attacks, compromise the security of the proposed software system and do damage to the system or obtain unauthorized access to functions and data with consequent criminal activities, such as theft of sensitive data (personal financial and health information, intellectual property), fraud, or sabotage.
Threat assessment	Understanding the nature of threats and the particular damage that a threat might inflict if a derived exploit were directed at specific types of system (e.g., the viability of a potential attack on a financial system aimed at obtaining customer nonpublic personal information to be used to steal money or financial instruments).	To determine the potential damage that a particular threat might inflict if successfully turned into an exploit and directed at existing vulnerabilities found in various categories of software systems.
Exploit determination	Determination of which threats have been, or could have been, developed into exploits through identifying specific exploits "in the zoo" or "in the wild," or by demonstrating the viability of exploits by creating them in a laboratory environment.*	To identify those threats (which are potential violations of security) that have been, or are likely to have been, turned into viable attacks or exploits.
Exploit assessment	Understanding which exploits might be beneficial (financially or otherwise) for attackers to use and which might not be worthwhile; also the determination of the technical viability of launching particular exploits into the wild.	To identify which exploits need to be protected against immediately and which can be addressed at a later date.
Weakness determination	Understanding secure coding issues from publications, courses, and webinars. Developing a list of software errors that might lead to less secure computer programs from publications, consultants (such as the common weaknesses listed by The MITRE Group), and associations such as OWASP and SANS. More details on these sources are provided in Appendix A.	To identify software mistakes that could contribute to vulnerabilities, which might be used by attackers to gain unauthorized access to a software system.

Table C.1   (continued)

Task Areas	Descriptions	Purposes
Weakness assessment	Evaluation of the degree to which specific software systems contain any of the weaknesses derived from the prior step using methods such as code reviews, peer reviews, and automated static analysis, and prioritization of those weaknesses with respect the damage that could be done were the weaknesses to become vulnerabilities and with respect to the likelihood of the weaknesses turning into vulnerabilities .	To determine to what extent the software errors (or failure to follow secure coding practices) might lead to exploitable vulnerabilities.
Vulnerability determination	Determination of which of the weaknesses, obtained from the prior step, have resulted in vulnerabilities; that is, where there are not any mitigating processes, methods, or tools that might have prevented the weaknesses from making the system vulnerable to attack. Other sources of vulnerabilities are the platforms and infrastructures on which the software resides, which are likely not under the control of those using such systems	To identify as many (as possible) of the vulnerabilities that are exhibited by the software and the other system components upon which the software depends in order to operate.
Vulnerability assessment	Assessment as to whether or not significant weaknesses are likely to become vulnerabilities, or whether there are other mitigation methods in use that would prevent a weakness from being exposed to potential attacks or inadvertent misuse.	To determine whether vulnerabilities that are found to exist within the software might be used by attackers to damage system or access and misuse sensitive information.
Risk identification	Creation of a list of risk categories and, within those categories, name security-related operational risks that emanate from the threats, exploits, weaknesses, and vulnerabilities obtained from the above steps. A risk occurs when an active exploit finds an existing vulnerability and results in a successful attack. Risks might go beyond operational security-related risks and include various financial risks, and other categories of risk, such as reputation risk.	To identify those risks that threaten the security of the software system so that they can be mitigated or eliminated if doing so is worthwhile.
Risk assessment	Calculation of the expected losses derived from each of the security risks identified above. The calculation involves the likelihood that a particular attack will be successful and the impact of the attack in terms of direct costs, inconvenience, and the like.	To determine the expected losses attributable to the various security risks so that the losses can be prioritized and resources allocated to reduce the risks in priority order.

Table C.1 (continued)

Task Areas	Descriptions	Purposes
Risk mitigation	Reduction or elimination of significant risks, as derived above, using a full range of tools and methods, including training and awareness, implementation of security tools (such as firewalls and intrusion detection and prevention systems), avoidance, deterrence, and the like.	To eliminate the possibility of certain attacks being successful or reduce the risk to an acceptable level using prevention, risk transfer, insurance, etc.
Risk management	Management and control of the entire risk process relating to software system security, as described above. The process consists of identifying and analyzing various threats, potential exploits, weaknesses, and vulnerabilities of security-critical systems and determining the likelihood of incidents and the expected losses from them.	To ensure that risks are identified and managed appropriately and that the security of software systems is maintained at an acceptable level while optimizing cost.

*The term "in the zoo" refers to exploits that have been developed but not released. The term "in the wild" refers to exploits that have been released to find vulnerable systems and attack them. Sometimes, if an exploit can be developed by a law-abiding group, such as a security vendor or an academic department, and it has not been seen "in the wild," it is presumed to be "in the zoo." However, it might be only a matter of time before it is released, given that there are worthwhile payoffs to those releasing the exploits detailed in Table C.2. This table describes and explains the tasks to produce complete sets of security requirements for software systems.

Once the security requirements have been decided upon and agreed to, by authorized persons, and adequate funding has been approved, the project team can embark on defining the security architecture and design of the software system, as described in Table C.3.

The design phase produces specifications from which developers build secure software systems. Table C.4 describes the various tasks that will produce sufficiently secure software systems based upon these specifications.

While it is expected that developers, their supervisors, and their project managers will check programs to ensure that they operate successfully (though not necessarily correctly), the testing of functionality is usually left to the Quality Assurance department. If addressed at all, testing of nonfunctional attributes such as security and performance is usually handed over to subject-matter experts, either internal to the organization or outside consultants. Nonfunctional security testing comprises the task areas shown in Table C.5.

Once the software system has been thoroughly tested to assure a high level of security, it is ready to be deployed into the production environment. However, even though a software system might be intrinsically secure, its secure operation is highly dependent on it being securely distributed and installed, as well as running in an environment that is secure, with security-oriented operating, support, and maintenance processes. The transition from development to production is a very sensitive step in the overall SSDLC and should follow the task areas that are described in Table C.6.

After a software system has been deployed, it is critical to its ongoing security that qualified staff has specific roles in order to maintain ongoing security throughout the operational life of the system. Table C.7 includes the task areas necessary for continued security in the operation, maintenance and support of the system.

Many publications on the Secure SSDLC stop at the operations phase. However, it is important that the end-of-life tasks are performed in a manner that will ensure that there are no security issues when the software systems are decommissioned and disposed of. These important tasks are described in Table C.8, along with reasons why it is so important to pay sufficient attention to this critical phase.

Involvement by Teams and Groups for Secure SSDLC Phases

We now look at each of the tasks within the various phases and indicate the levels of involvement in those tasks by various teams or groups within a typical organization. The entries in these tables (Tables C.9 to C.16) and in the summary

Table C.2
Security-Related Tasks for the Requirements Phase

Task Areas	Descriptions	Purposes
Identification of a complete set of security requirements	Obtaining security requirements from stakeholders, including: end users, information security team, legal and compliance, internal audit, external auditors, security consultants, human resources, project management office, systems architects, application developers, testers, quality assurance, software assurance, implementers, operators, support functions, system and security administrative functions.	To ensure that the security requirements of all anticipated stakeholders are gathered, so that they will be analyzed, prioritized and included in the overall set of system requirements.
Analysis and prioritization of security requirements	Prioritization of security-related requirements by authorized decision maker(s).	To ensure that appropriate security-related requirements are included in the system architecture and design.
Evaluation of security risks and consequences	Education of business units so that they understand the risks and consequences of security breaches.	For educational purposes.
Information asset value analysis	Understanding and determining the value of various security-related assets that need to be protected.	To determine what level of protection is needed and choose appropriate technical and non-technical security controls so that they can be incorporated into the security architecture and design of software systems.
Agreement upon security goals	Discussion of key security requirements and goals relating to software-system confidentiality, integrity, availability, non-repudiation, auditability, and the like, and agreement upon how to approach each of these factors and what the desired outcome might be.	To establish design goals related to key security requirements.
Legal and regulatory requirements	Review and understanding of legal and regulatory requirements and translate them into security policy.	To incorporate security requirements and relevant laws and regulations into the design of software-intensive systems.
Determination of potential threats	Understanding the form and source of security threats that might lead to actual exploits based on those determined in the risk analysis and control phase.	To be able to assess risks associated with specific threats and exploits so that mitigating security controls can be prioritized.

Table C.2 (continued)

Task Areas	Descriptions	Purposes
Analysis of potentially damaging activities	Understanding the range of potential malicious and harmful activities based on those determined in the risk analysis and control phase.	To determine activities that might impact the security of systems so that appropriate controls can be put in place.
Analysis of high-level vulnerabilities	Understanding the classes of vulnerabilities and their impact if used to compromise the security of systems based on those vulnerabilities determined in the risk analysis and control phase.	To be able to address significant vulnerabilities by eliminating (through patching) or bypassing them, or by implementing protective capabilities.
Future business goals	Determination as to whether or not the proposed system is extensible to the degree that it can be built to account for likely future requirements.	To try to design the proposed system so as to be able to incorporate anticipated requirements, such as the need to recreate historical data from replaced systems.
Business/information technology/operations plan	Understanding the operational and information technology context where the new system is expected to run, including documentation of the responsibilities of various parties.	To ensure that the new software system will operate in the expected context and that it will meet operational requirements.
Gaps or deficiencies in the current security program	Reviewing the security program in order to determining potential changes to existing security policy, procedures, architecture, standards, processes, and technologies so that any deficiencies that arise from a new system can be incorporated into the policy and procedures.	To understand deficiencies in the current security program in order to address them separately or via the system-level security architecture.
Population of users and their requirements	Determination of who will be using the new system and for what purposes, as well as when the users will be expected to use the system (so as to account for peak loads, normal usage, time of day, day of week, etc.).	To generate a full set of user requirements in order to confirm that the scope and design of the system will satisfy those requirements, and if not, to develop mitigation strategies.
Customer and partner interface requirements	Understanding the considerations and constraints that might arise with the new architecture and design of the system emanating from new system requirements and system integration.	To ensure that the older and new systems interoperate when they are interconnected.
Project timeframe	Getting approval for the project work plan, including timeframe, milestones and deliverables.	To ensure that there is a basis for acquiring and assigning specific levels of human, technical and support resources, and for applying those resources in order to accomplish given deliverables at the assigned milestones.

Table C.2 (continued)

Task Areas	Descriptions	Purposes
Prioritized security solution requirements	Documentation of high-level security requirements as they relate to business and technology for use in the architecture and design phases.	To ensure that high-level security requirements are introduced into the development process at as early a stage as is feasible and reasonable.
Proposed costs subject to budget constraints	Creation of a dedicated security budget to cover the development, testing, deployment, integration, operation, and decommissioning of the system with specific focus on such activities as ongoing vulnerability monitoring and assessments.	To make sure that all relevant and required costs relating to security are included within the budget so as to ensure that adequate resources are provided to attain a desired security level and to avoid having to go back for additional funding later in the project, especially as it becomes more expensive to introduce security later in the life cycle.
Approval of security requirements and associated budget	Gaining buy-in from decision-makers and getting their approval and funding for critical security requirements.	To formalize the security requirements approval process and obtain formal sign-offs of suggested security requirements and associated proposed budgets in order to avoid subsequent budget issues.
Decision as to whether systems are to be built or bought	Determining which security tasks and functions can be better assigned to internal staff, third parties or a combination of resources, depending on their respective abilities to execute security architecture, design, testing, implementation, operation, upgrading, replacement, decommissioning, and disposal.	To determine the appropriate assignment of security tasks and functions.
Independent security-oriented review of requirements	Engagement of third parties to perform a security assessment of the requirements.	To determine if any security requirements need to be added, deleted or modified.

Table C.3

Security-Related Tasks for the Architecture and Design Phase

Task Areas	Descriptions	Purposes
Creation of system-level security architecture	Determination of possible security solutions needed to achieve the security requirements and specified goals.	To document, at a high level, specific security requirements and technical means which those requirements might be met.
Determination of technical security controls	Listing the requisite technical controls (such as access control, authentication, encryption, etc.) that lead into an overall system design.	To document technical controls needed to meet security requirements.
Determination of non-technical security controls	Establishing the requisite nontechnical controls (i.e., security-related processes and procedures, such as for user registration and authentication).	To document the non-technical controls needed to meet security requirements.
Performance of architecture walkthrough	Discovery of any deficiencies in the architecture and controls so that they might be corrected at this early stage at relatively little cost in time and resources.	To help to ensure that proposed technical and nontechnical controls will meet security and business requirements effectively.
Create system-level security design	Determination of specific techniques (i.e., methods and tools) that will be used to achieve the requisite technical and nontechnical controls.	To document specific technical and nontechnical controls that will be used for the software system.
Establish top-level technical security design	Listing of tools and products contributing to the technical solution indicating how they will interoperate with other technical security elements.	To document the necessary technical security elements and how they will be implemented to achieve various security requirements.
Establish top-level non-technical security design	Defining security processes and procedures and their interrelationships, such as security Concept of Operations (CONOPS).	To document high-level security processes and procedures.
Create detailed technical security design	Specification of vendor tools, configuration settings, etc.	To add specific detail to the top-level technical security design.
Create detailed non-technical security design	Specification of particular steps for each process in the CONOPS, including specific technical security standards and configuration procedures.	To add specific detail to the top-level nontechnical security design
Analyze costs and benefits for various design elements	Evaluation of various cost-benefit comparisons for design alternatives and prioritization of these options.	To evaluate various design scenerios and make cost-benefit tradeoffs prior to investirg in the detailed design.

Table C.3 (continued)

Task Areas	Descriptions	Purposes
Review technical design at the application and infrastructure levels	Understanding vendor/tool integration, performance issues, and products, which might involve the laboratory testing of the security infrastructure.	To determine whether any design changes should be made to correct flaws and to prioritize those changes, and to make product selections based on architecture requirements.
Creation of secure systems	Making developers aware of general security issues, development-related security issues, such as secure development environments, and security-related tools available to developers.	To increase the probability of developers taking security seriously and incorporating various security methods into their development processes.
Provide training and awareness programs for end-users, administrators, operators, business managers, executive management	Development of training and awareness program, including documentation and other media (e.g., videos) for various courses depending upon the constituencies; introduction of courses, certifications, and the like. Resources should cover system documentation and operation, help systems and procedures, installation procedures, legal and regulatory requirements (e.g., privacy), etc.	To ensure that each area that touches security-critical software systems has a good understanding of the proper use of security and privacy features, activities that should not be done, issues relating to legal and regulatory requirements, and reporting of unusual and potentially infringing activities.
Create security test plan	Specification of aspects of the software system that are to be tested—such characteristics include security, performance, integrity, availability, resiliency, interoperability, portability, and the like. Functional security testing is used to determine that the software system does not allow unauthorized and unacceptable activities to take place, and includes assuring that the functional requirements of the software will be met.	To document methods of testing functional and nonfunctional features of the software system.
Add or update information security policy, standards, and/ or procedures as necessary	Evaluation and updating of existing documentation on policies, standards, procedures, and training in the light of upcoming software systems, particularly those using new, untested, or vulnerable technologies.	To ensure that all security-related documentation has been brought up-to-date with respect to the new software systems.
Create plan for mitigation of key application and infrastructure vulnerabilities in third-party software systems	Population of a database of vulnerabilities obtained from various sources (such as software product vendors, security product vendors, security-related websites, etc). Development of internal procedures for applying fixes and monitoring and reporting the status of fixes against the entire population of vulnerable systems.	To document vulnerabilities for applications, middleware, system software and infrastructure, both internally-developed and purchased off-the-shelf. To prioritize available fixes (patches) and to ensure that those representing the greatest risk are applied in timely manner.

Table C.3 (continued)

Task Areas	Descriptions	Purposes
Determine security requirements for development, test and operational environments	Determination of the processes, procedures, and technology that assure a reasonable and suitable level of security (e.g., version controls, access controls, segregating development, testing and production environments, scrubbing data used in the development and testing phases so as not to expose sensitive information, etc.).	To ensure that, as the software passes through its various development phases, the level of security is high enough that the software and corresponding data will not be compromised in its travels.
Independent security-oriented review of architecture and design of software system	Engagement of third parties to perform security assessments of the architecture and design of the software system.	To determine if there are any security aspects of the architecture and design that need to be added, deleted or modified.

Table C.4

Security-Related Tasks for the Development Phase

Task Areas	Descriptions	Purposes
Set up secure development environment	Implementation of processes, procedures and technologies as specified in the Design Phase.	To ensure that access and use of source code, component configurations, documentation, and the like are controlled and monitored, that deviations are reported, and that planned responses are appropriate.
Security-related training for development teams	Provision of training with respect to secure coding practices and security-related middleware, so that developers become aware of general security issues, other development-related security issues (e.g., secure development environment, see above), and security-related tools.	To ensure that developers are adequately trained in secure coding practices, as well as being made aware of other relevant security-related matters.
Security-related training for infrastructure teams	Provision of training with respect to the configuration and installation of middleware and other system software and services, particularly as they relate to security.	To ensure that infrastructure teams are adequately trained in security aspects of configuring and operating platforms and infrastructures upon which the developed software will run.
Develop application-level security components	Creation of security-related components and services for security-critical software systems.	To develop applications that meet security-related requirements and specifications.
Implement secure infrastructure in support of application software	Establishment of secure infrastructures and environments in support of application development and testing activities.	To provide a secure environment for development and testing activities.
Set up vulnerability tracking process	Implementation of a vulnerability tracking database, possibly integrated with quality assurance or error tracking systems, with the ability to record, prioritize, sort, monitor, report, and indicate responses or actions taken.	To ensure that those vulnerabilities, which are identified, recorded, prioritized, resolved, and marked as fixed.
Plan security testing	Development of security test plans for prospective, current and scheduled future versions of the software system.	To ensure that security-related testing is not neglected and is performed as required.
Plan unit, integration and regression testing of security services	Development of plans for testing security-related services (such as external authentication services, external encryption services, network security services, etc.).	To ensure that security-related services, invoked by developers, are working properly.

Table C.4 (continued)

Task Areas	Descriptions	Purposes
Plan code review	Development of plans to perform code reviews of the application programs.	To determine whether there are any security issues with respect to the program code and suggest how they might be addressed in priority order.
Independent security-oriented review of development processes and products	Engagement of third parties to perform security assessments of the development processes and procedures and to perform code reviews.	To determine if there are any security issues relating to the development of the software system.

Table C.5

Security-Related Tasks for the Testing Phase

Task Areas	Descriptions	Purposes
Code review	Review and testing of application code to verify that it meets security requirements and specifications using various tools and procedures.	To determine, prioritize, and address code-level security issues.
Testing of configuration procedures	Review of configuration procedures to verify that they are usable by administrators for setting up networks and system components to meet security standards.	To determine if there are any procedural issues in need of resolution.
System testing	Testing to determine whether the technical components of the software system are functioning together and conform to architecture and design specifications.	To determine if there are any system-level and/or system-integration issues in need of resolution.
Performance/load testing	Testing to determine whether systems performance and production load requirements meet prespecified levels with security components and features operating.	To determine if there are any security-related performance issues in need of resolution.
Usability testing	Testing to determine whether systems meet end-user usability and ergonomic needs.	To determine if there are any usability issues in need of resolution.
Independent vulnerability assessments	Engagement of third parties to perform vulnerability assessments.	To determine if there are any application-level, platform, or infrastructure vulnerabilities that need to be addressed.
Independent security-oriented review of testing	Engagement of third parties to perform a security assessment of the testing procedures and environment.	To determine if there are any security testing issues that need to be resolved.

Table C.6
Security-Related Tasks for the Deployment Phase

Task Areas	Descriptions	Purposes
Pilot deployment	Determination of issues related to all areas of software system planning, design, architecture, design, testing, implementation, and decommissioning.	To allow development and architecture teams to make adjustments to operational processes, systems architecture, design, and programs prior to release.
Transition from final testing to operations	Assurance that the final system, just prior to deployment, addresses adequate levels of integrity and that appropriate change management procedures, encryption of transmissions, interim access authorizations, and so on, are in place.	To ensure that the transition process conforms to good security practices.
Comparison of system files	Assurance that system files of the system, which is about to be launched, are compared to the original system files and that they match, thereby assuring the authenticity of the files.	To ensure the authenticity of system files, which are to be moved into the operations environment.
Comparison of data files	Assurance that data files of the system is about to be launched are compatible with the original data files and that they match, thereby assuring the authenticity of the data.	To ensure the authenticity of data files that are to be moved into the operations environment.
Training and awareness	Assurance that all appropriate stakeholders (end users, administrative personnel, technical support, operations staff, etc.) are trained sufficiently and are aware of the security functions and attributes of the system.	To ensure that those who will use and support the new system are sufficiently familiar with the security aspects of the system as required for their properly performing their respective functions.
Full-scale deployment	Assurance that the infrastructure is secure prior to the installation of the new system.	To avoid deploying the system into a vulnerable environment.
Independent security-oriented review of deployment	Engagement of third parties to perform a security assessment during deployment, as well as for predeployment and post-deployment.	To determine whether there are any security-related deployment issues, which need to be resolved prior to deployment.

Table C.7

Security-Related Tasks for the Operations, Maintenance, and Support Phase

Task Areas	Descriptions	Purposes
Testing and migrating to new software version	Analysis of new versions of software to determine which vulnerability conditions may have been resolved, which vulnerabilities have not been resolved, and which new vulnerability conditions might have been introduced or have arisen over time.	To determine the vulnerabilities of current and replacement software versions in order to arrive at a revised list of vulnerabilities to be resolved.
Risk reviews	New risk assessments if business risk conditions change materially.	To determine whether business risks have changed and therefore need to be reviewed and mitigated.
Security awareness program	Maintenance of security awareness program to account for changes in threat and vulnerability environments and addition of new customers, employees, consultants, etc.	To keep security awareness and training programs up-to-date with respect to content and population to be trained, as well as to reflect changes in system features and the like.
Vulnerability assessments	Periodic vulnerability assessments to account for changes in systems, technologies, and personnel.	To ensure that new and changing vulnerabilities are recognized and that a program for their mitigation is developed and enacted.
Auditing, logging, monitoring, archiving	Implementation of appropriate logging and archiving procedures and periodic reviews of the logs.	To determine whether any anomalous activities have occurred or are taking place and, if so, to take action to halt activities, prevent future activities, and take appropriate steps against perpetrators.
Independent security-oriented review of operations and support	Engagement of third parties to perform security assessments of post-deployment operation, maintenance, and support.	To determine whether there are any security-related operational issues that need to be resolved.

Table C.8
Security-Related Tasks for the Decommissioning and Disposal Phase

Task Areas	Descriptions	Purposes
Integrity of the old and new systems during transition of the old system to a decommissioned state	Assurance that adequate plans are in place for decommissioning and disposing of the software and any media or hardware on which the old system or its components depended and that the transition to the new system is not compromised.	To avoid problems with inadequate disposal of data and systems, including software, hardware, and media, from the old system and that the integrity of the systems and data are maintained throughout the transition.
Transition from operational to decommissioned state	Assurance that the new system (or revised version of the old system) can operate on its own with respect to security-related functions before decommissioning the old system and that operators can immediately switch back to the old system during the initial period when the new system takes over if any problems are detected.	To avoid the possibility that neither the old system nor the new system is available for any period of time, and to ensure that the security functions of the systems remain effective at all times prior, during, and after the transition.
Comparison of system files	Assurance that system files of the new system, which is about to replace the old system, are compared to the original system files and that they match, thereby assuring the authenticity and integrity of the files.	To ensure the authenticity and integrity of system files to be used by the new system.
Disposal of application and system files	Assurance that all valuable application and system files (particularly those containing sensitive intellectual property and nonpublic personal information) that are no longer needed are either removed from magnetic media or that the media on which the data reside, including paper hard copies, are physically destroyed.	To ensure that the intellectual property and nonpublic personal information that may be contained in the application and system files will not be obtained by those unauthorized to access the application and system files, especially those who might use the information for nefarious purposes.
Comparison of data files	Assurance that data files of the new system about to replace the old system are compatible with the original data files and that they match, thereby assuring the authenticity and integrity of the data.	To ensure the authenticity and integrity of data files that are to be used by the new system and, where possible, that the data files are backward compatible; meaning that the new system can still read and operate upon the data files created and used by the old system.

Table C.8 (continued)

Task Areas	Descriptions	Purposes
Disposal of data files and media	Assurance that all sensitive data (such as intellectual property and nonpublic personal information) that is no longer needed is either removed from magnetic media or that the media on which the data reside, including paper hard copies, are physically destroyed.	To ensure that the data will not be obtained by those unauthorized to access the data, especially those who might use the data for fraudulent or other nefarious activities.
Training and awareness	Assurance that all appropriate stakeholders, such as end users, administrative personnel, technical support, and operations staff, are trained and aware of the security functions and attributes of the new system and have in hand a set of procedures for the transition.	To ensure that those who will use and support the new system are sufficiently familiar with the security aspects of the new system versus the old system as required for performing their respective functions.
Full-scale decommissioning	Assurance that the existing infrastructure is secure prior to transitioning from the old system to new system, possibly including a period of parallel operation.	To avoid creating a vulnerable environment during the decommissioning of the old system and installation of the new system, whether or not there is a period of parallel operation.
Independent security-oriented review of decommissioning and disposal	Engagement of third parties to perform security assessments of the transition predecommissioning and disposal and post-decommissioning and disposal.	To determine whether there are any security-related decommissioning and disposal issues that need to be resolved.

Table C.9
Levels of Involvement of Teams/Groups for the Risk Analysis and Control Phase

Task Areas	Level of Team/Group Involvement							
	Security & Risk Teams	Application Development Team	Infrastructure Team	Project Managers	Business Unit Managers	Operations Managers	Internal & External Auditors	Executive Management
Threat determination	High	Moderate	Moderate	Moderate	Moderate	Low	High	Moderate
Threat assessment	High	Moderate	Moderate	Moderate	Moderate	Low	High	Moderate
Exploit determination	High	Moderate	Moderate	Moderate	Moderate	Low	High	Moderate
Exploit assessment	High	Moderate	Moderate	Moderate	Moderate	Low	High	Moderate
Weakness determination	High	Moderate to High	Moderate to High	Moderate	Low	Low	High	Low
Weakness assessment	High	Moderate to High	Moderate to High	Moderate	Low	Low	High	Low
Vulnerability determination	High	Moderate to High	Moderate to High	Moderate	Low	Low	High	Low
Vulnerability assessment	High	Moderate to High	Moderate to High	Moderate	Low	Low	High	Low
Risk identification	High	Moderate to High	Moderate to High	Moderate	Very high	Moderate	High	Very high
Risk assessment	Very high	Moderate to High	Moderate to High	Moderate	High	Moderate	Very high	High
Risk mitigation	High	Moderate to high	Moderate to high	Moderate	High	Moderate	High	High
Risk management	Very High	Moderate to high	Moderate to high	Moderate	High	Moderate	Very high	Very high
Overall	High	Moderate to high	Moderate to high	Moderate	Moderate	Low to moderate	High	Moderate

Table C.10
Levels of Involvement of Teams/Groups for the Requirements Phase

Task Areas	Level of Team/Group Involvement							
	Security Team	Application Development Team	Infrastructure Team	Project Managers	Business Unit Managers	Operations Managers	Internal & External Auditors	Executive Management
Identification of a *complete* set of security requirements	High	High	Moderate	Moderate	High	Moderate	Moderate to High	Moderate
Analysis and prioritization of security requirements	High	High	Moderate	Moderate	High	Moderate	Moderate	Moderate
Evaluation of security risks and consequences	High	Low	Low	Moderate	High	Moderate	High	Moderate
Information asset value analysis	High	Low	Low	Moderate	High	Moderate	High	High
Agreement upon security goals	High	Low	Low	Moderate	High	Moderate	High	High
Legal and regulatory requirements	High	Very Low	Very Low	Low	High	Moderate	High	High
Determination of potential threats	High	Low	Low	Moderate	High	Moderate	High	Moderate
Analysis of potentially damaging activities	High	Low	Low	Moderate	High	Moderate to high	High	Low
Analysis of high-level vulnerabilities	High	Low	Low	Low to moderate	Moderate	Moderate	High	Low
Review of future business goals	Moderate	Low	Low to Moderate	Low	High	Low to Moderate	Moderate	High
Plan business/IT operations	Moderate to high	Moderate to high	Moderate to high	Moderate	High	High	Moderate to high	Low
Identification of gaps in security program	High	Low	Low	Low to Moderate	Moderate to high	Low to moderate	High	Moderate
Listing of users and their requirements	Moderate	High	Moderate	High	High	High	High	Low to moderate

Table C.10 (continued)

Task Areas	Level of Team/Group Involvement							
	Security Team	Application Development Team	Infrastructure Team	Project Managers	Business Unit Managers	Operations Managers	Internal & External Auditors	Executive Management
Customer and partner interface requirements	High	High	High	Moderate to high	High	High	High	Moderate
Project timeframe	High	High	High	High	High	High	Moderate to high	Moderate to high
Prioritized security solution requirements	High	High	High	High	High	High	Moderate to high	Moderate to high
Propose costs subject to budget constraints	High	High	High	High	High	High	Moderate to high	Moderate to high
Approve security requirements and associated budget	High	Low	Low	Low	Low	Low	High	High
Decide whether systems are to be built or bought	Moderate	Low	Low	Low	Low	High	Moderate	High
Independent security review of requirements	High	Low	Low	Low	Low	High	High	High
Overall	High	Low to moderate	Low to moderate	Low to moderate	Moderate to high	Moderate to high	High	Moderate to high

(Table C.17) are normative. That is to say, many of the teams or groups might not exist in typical small and medium-sized businesses or in some large entities. Also, it is the exception, rather than the rule, for groups so designated to have the level of involvement shown. While some surveys, such as those conducted and reported as part of the Building Security in Maturity Model (BSIMM) effort [1], show relatively high levels of involvement by team members, the companies participating in the BSIMM exercise are largely self-selecting. The general picture in both public and private sectors is not nearly as encouraging.

In Table C.9, we show the degree of involvement of various areas in the risk analysis and control phase. For the most part, risk analysis focuses on the business perspective, preferably with considerable support from the information security department and risk area and oversight by internal and external auditors. The results from the risk analysis become a major source for the subsequent requirements development phase. This approach is far from ideal because if the software development and operations teams were to be more involved than usual, they would better understand the concerns of the business areas and account for such concerns in the design, development, and implementation of software-intensive systems. This disconnect leads to developers not assuming responsibility for ensuring that their software addresses business management risk concerns—this transfer of responsibility is described in greater detail in the section on *technical debt* in Chapter 11. In Table C.9, and in the subsequent tables, we indicate the ideal levels of involvement. It should also be noted that the entries in the following tables are highly subjective based on personal experience and judgment, and that other situations may call for different levels of involvement.

It is usual for the development team and the infrastructure team to be heavily involved in the requirements phase, as shown in Table C.10 preferably with a high level of involvement by the business units for which the software system is being created. Often, the security requirements are given short shrift, as evidenced by the readily apparent lack of security and privacy capabilities of many web applications, in particular, but also of production systems. There should be heavy involvement of the security team and auditors to ensure that the requirements are adequate with respect to security.

The security-related tasks in the design and architecture phase, shown in Table C.11, should depend heavily on the security team, as well as on the application development and infrastructure teams and, to the extent possible, on the auditors. Business units, where the end-users generally reside, are somewhat less involved because most of the work in this phase is technical.

As Table C.12 indicates, the development or coding phase is mostly dependent on the technical teams, with strong oversight by auditors. The business side is less involved. Similar involvement carries through to the testing phase, depicted in Table C.13, for the same reasons.

Table C.11
Levels of Involvement of Teams/Groups for the Architecture and Design Phase

Task Areas	Level of Team/Group Involvement							
	Security Team	Application Development Team	Infrastructure Team	Project Managers	Business Unit Managers	Operations Managers	Internal & External Auditors	Executive Management
Create system-level security architecture	High	High	High	Moderate to high	Low to moderate	Moderate	Moderate	Low
Determine technical security controls	High	High	High	Moderate to high	Low to moderate	Moderate	Moderate	Low
Determine nontechnical security controls	High	Moderate	Moderate	Moderate	Moderate to high	Moderate	Moderate	Moderate
Perform architecture walkthrough	High	High	High	High	Low to moderate	Moderate to high	High	Low
Create system-level security design	High	High	High	Moderate to high	Low to moderate	Moderate	Moderate	Low
Establish top-level technical security design	High	High	High	Moderate to high	Low to moderate	Moderate	Moderate	Low
Establish top-level nontechnical security design	High	Moderate	Moderate	Moderate	Moderate to high	Moderate	Moderate	Moderate
Create detailed technical security design	High	High	High	Moderate to high	Low to moderate	Moderate	Moderate	Low
Create detailed nontechnical security design	High	Moderate	Moderate	Moderate	Moderate to high	Moderate	Moderate	Moderate
Analyze costs and benefits for various security design elements	High	Moderate	Moderate	Moderate	High	Moderate	Moderate to high	Moderate
Review technical security design at the application and infrastructure levels	High	High	High	Moderate to high	Low to moderate	Moderate	High	Moderate to high

Table C.11 (continued)

Task Areas	Level of Team/Group Involvement							
	Security Team	Application Development Team	Infrastructure Team	Project Managers	Business Unit Managers	Operations Managers	Internal & External Auditors	Executive Management
Creation of secure systems	High	High	High	Moderate to high	Low to moderate	Moderate	Moderate	Low
Provide security training and awareness programs for end-users, administrators, operators, business managers, and executive management	High	Moderate	Moderate	Moderate	Moderate	Moderate	Moderate	Moderate
Create functional and nonfunctional security test plan	High	High	High	High	High	Moderate	High	Low to moderate
Add or update information security policy, standards, and/or procedures as necessary	High	Moderate	Moderate	Moderate	Moderate	Moderate to high	High	Moderate
Create plan for mitigation of key application and infrastructure vulnerabilities	High	Moderate to high	Moderate to high	Moderate	Moderate	Moderate	High	Low
Determine security requirements for development, test and operational environments	High	High	High	High	High	Low	High	Low
Independent security review of architecture and design of software system	High	Moderate to high	Moderate to high	High	Low	Low	High	Low
Overall	High	High	High	Moderate to high	Moderate to high	Moderate	High	Low to moderate

Table C.12
Levels of Involvement of Teams/Groups for the Development Phase

Task Areas	Level of Team/Group Involvement							
	Security Team	Application Development Team	Infrastructure Team	Project Managers	Business Unit Managers	Operations Managers	Internal & External Auditors	Executive Management
Set up secure environment for development activities	High	High	High	High	Moderate	Moderate	High	Low to moderate
Security-related training for development teams	High	High	Moderate	Moderate to high	Low	Moderate	High	Moderate
Security-related training for infrastructure teams	High	Moderate	High	Moderate	Low	Low to moderate	Moderate to high	Low to moderate
Develop application-level security components	High	High	Moderate	High	Low to moderate	Low to moderate	High	Low
Implement secure infrastructure in support of application software	High	Moderate	High	High	Low	Low to moderate	Moderate to high	Low
Set up vulnerability tracking process	High	High	High	High	Low	Moderate	High	Low
Plan security testing	High	High	High	High	Moderate to high	High	High	Moderate
Unit, integration, and regression testing of security services	High	High	High	High	Moderate to high	High	High	Moderate
Plan code review	High	High	High	High	Low	Low	High	Low
Independent review of development processes and products	High	High	High	High	Moderate	Moderate	High	Moderate
Overall	High	High	High	High	Moderate	Moderate	High	Low to moderate

Table C.13
Levels of Involvement of Teams/Groups for the Testing Phase

Task Areas	Level of Team/Group Involvement							
	Security Team	Application Development Team	Infrastructure Team	Project Managers	Business Unit/QA Managers	Operations Managers	Internal & External Auditors	Executive Management
Code review	High	High	High	High	Low	Low	High	Low
Testing configuration procedures	High	Low	High	High	Low	High	High	Low
System testing	High	High	High	High	Low	Low	High	Low
Performance/load testing	High	High	High	High	High	Moderate	High	Low
Usability testing	High	High	High	High	High	Moderate	High	Low
Independent vulnerability assessments	High	High	High	High	Moderate	Low	High	Low
Independent review of security aspects of testing	High	High	High	High	Moderate	Low	High	Low
Overall	High	High	High	High	Moderate	Low	High	Low

Table C.14
Levels of Involvement of Teams/Groups for the Deployment Phase

Task Areas	Level of Team/Group Involvement							
	Security Team	Application Development Team	Infrastructure Team	Project and Change Managers	Business Unit Managers	Operations Managers	Internal & External Auditors	Executive Management
Pilot deployment	High	High	High	High	Moderate	High	Moderate	Low
Transition from final testing to operations	High	High	High	High	Moderate	High	High	Low
Comparing system files	High	Moderate	High	High	Low	Moderate	Moderate to high	Low
Comparing data files	High	High	Moderate	High	Low	Low	High	Low
Training and awareness	High	Moderate	Moderate	Moderate to high	High	Moderate	High	Moderate to high
Full-scale deployment	High	High	High	High	Moderate	High	High	Moderate to high
Independent review of security aspects of deployment	High	High	High	High	Moderate	High	High	Moderate
Overall	High	Moderate to high	Moderate to high	High	Moderate	Moderate	High	Moderate

Table C.15
Levels of Involvement of Teams/Groups for the Operations, Maintenance, and Support Phase

Task Areas	Level of Team/Group Involvement							
	Security Team	Application Development Team	Infrastructure Team	Project Managers	Business Unit Managers	Operations Managers	Internal & External Auditors	Executive Management
Testing and migrating to new software version	High	High	High	High	Moderate to High	Moderate to High	High	Low
Risk reviews	High	Low	Low	Low	High	Low	High	Moderate to high
Security awareness program	High	Moderate	Moderate	Moderate	High	Low	High	Moderate to high
Vulnerability assessments	High	Moderate	Moderate	Moderate	Moderate	Moderate	High	Low
Auditing, logging, monitoring, archiving	High	Low to moderate	Low to moderate	Low to moderate	Moderate to high	High	High	Moderate
Overall	High	Moderate	Moderate	Moderate	Moderate to high	Moderate	High	Moderate

Table C.16
Levels of Involvement of Teams/Groups for the Decommissioning and Disposal Phase

Task Areas	Level of Team/Group Involvement								
	Security Team	Application Development Team	Infrastructure Team	Project Managers	Business Unit Managers	Operations Managers	Internal & External Auditors	Executive Management
Maintaining of integrity of old and new systems during decommissioning	High	Moderate	Moderate	Moderate	Moderate	High	High	Low
Transition from operational to decommissioned state	High	Low	Low	Low	Low	Moderate to high	High	Low
Comparison of system files	High	Moderate	High	High	Low	Moderate	Moderate to high	Low
Disposal of application and system files	High	Low to moderate	Low to moderate	Low	Low	High	High	Low
Comparison of data files	High	Low	Low	Low	Low	High	High	Low
Disposal of data files and media	High	Low	Low	Low	Low	High	High	Low
Training and awareness for disposal	High	Low	Low	Low	Low	High	High	Low
Full-scale decommissioning	High	Low	Low	Low	Low	High	High	Low
Independent review of decommissioning	High	Low	Low	Low	Low	High	High	Low
Overall	High	Low	Low	Low	Low	High	High	Low

Table C.17

Summary of Levels of Involvement by Phase

Phase	Level of Team/Group Involvement							
	Security Team	Application Development Team	Infrastructure Team	Project Managers	Business Unit Managers	Operations Managers	Internal and External Auditors	Executive Management
Risk analysis and control	High	Moderate to high	Moderate to high	Moderate	Moderate	Low to moderate	High	Moderate
Requirements	High	Low to moderate	Low to moderate	Low to moderate	Moderate to high	Moderate to high	High	Moderate to high
Architecture and Design	High	High	High	Moderate to high	Moderate to high	Moderate	High	Low to moderate
Development	High	High	High	High	Moderate	Moderate	High	Low to moderate
Testing	High	High	High	High	Moderate	Low	High	Low
Deployment	High	Moderate to high	Moderate to high	High	Moderate	Moderate	High	Moderate
Operations and Maintenance	High	Moderate	Moderate	Moderate	Moderate to high	Moderate	High	Moderate
Decommissioning and disposal	High	Low	Low	Low	Low	High	High	Low
Overall Levels for the life cycle	High	Moderate to high	Moderate to high	Moderate to high	Moderate to high	Moderate	High	Low to moderate

The weighting of effort for the deployment phase, as shown in Table C.14, shifts to the operational areas, although there is still significant support required from the security team and technical teams. As shown in Table C.15, the same applies when the system moves into production or operation.

Decommissioning involves the operational areas with oversight by security and audit, as shown in Table C.14.

The summaries of Tables C.9 to C.16, which provide levels of involvement for tasks within phases, are combined into Table C.17, which shows the overall involvement of the various groups and teams in the security aspects of each phase of the life cycle. As expected, the main overall participants are the security and audit team members, but the application development team and the infrastructure team follow close behind. For the effort to be successful, participants must be fully engaged and perform their tasks diligently.

A Note on Sources

The entries in the above tables cannot all be attributed to specific references because they represent an amalgamation of concepts drawn from various sources, many of which are given in the End Notes section of Chapter 9. Most of the tables in this appendix are derived from McBride [2], although their form and content varies considerably from those in [2]. For example, the risk analysis and control phase as well as decommissioning and disposal phase are not included in [2].

Endnotes

[1]   See www.bsimm.com, accessed August 15, 2012.

[2]   McBride, P., and E. P. Moser, *Secure System Development Life Cycle (SDLC) Building Security Into Your System—Not Bolting It On After the Damage is Done,* Atlanta, GA: METASeS, 2000. Available at http://lazarusalliance.com/horsewiki/images/3/38/Secure-System-Development-Life-Cycle.pdf. Accessed August 5, 2012.

Appendix D
Safety-Related Tasks in the Safe SSDLC

As in Appendix C, which focused on the Secure SSDLC, Tables D.1 through D8, show tasks for each phase of the safe software system development life cycle (SSDLC), describe them, give their purposes. Table D.9 shows the levels of involvement of various players for each phase.

The following tables should to be used for guidance purposes only. These tables should be helpful for those responsible for the safety of mission-critical software systems and they offer a checklist to assist in the process of ensuring that adequate attention has been applied to safety considerations throughout the life cycle. The checklist should have even more value to those who concentrate on security-critical systems, as these individuals will likely not be very familiar with safety-critical software system requirements and methods for building them.

Task Areas for Safe SSDLC Phases

We now show the tasks, descriptions, and purposes for each of the following phases of the Safe SSDLC:

- Hazard risk analysis and control;
- Safety software requirements;
- Safety architecture and design;
- Development (building/configuration/integration);
- Verification, validation and certification;

- Deployment;

- Operations, maintenance and support;

- Decommissioning and safe disposal.

Tables D.1 through D.8 provide guidance for those tasked with ensuring that safety is built into software systems and that those systems meet particular standards necessary for certification. The main differences between the phases presented here and those phases described in Appendix C are (1) replacement of risk analysis and control relating to protecting information assets with hazard risk analysis and control relating to damage that a malfunction or failure might do to humans, other creatures, or the environment; and (2) the substitution of verification, validation, and certification phase for the testing phase. The other phases of the Safe SSDLC appear to be similar in name to the phases in the Secure SSDLC, but significant differences will be highlighted as we go through the phases one at a time. For example, the disposal of a safety-critical software systems and data can take a very different form in comparison with the disposal of security-critical software systems and data.

As mentioned in Appendix C, there is some degree of asymmetry between the Secure and Safe SSDLCs, particularly with respect to a strong recommendation to have software-security subject-matter expertise available to the software-safety project team. This is in contrast to not usually needing to have safe-software experts involved in the Secure SSDLC.

As with the Secure SSDLC, the first phase of the Safe SSDLC is also a risk analysis and control phase. However, in the latter case, it is the risk that a malfunction or failure will cause physical harm to humans and/or the environment. In the former case, it is generally a matter of determining the likelihood that a security-critical software system will be penetrated and sensitive data compromised. While it is true that the result of a breach or failure of a security-critical software system can result in harm, the damage is usually in the form of fraud against financial assets, theft of intellectual property, and the release of nonpublic personal information. These latter consequences can be extremely distressing and might lead indirectly to physical harm. However, financial losses are generally considered not to be as serious as physical harm, even though the former can usually be estimated fairly accurately and the latter are generally rough estimates because of the difficulty in assessing the value of a life, the cost of an injury, and the financial impact of an environmental disaster.

In Table D.1 we show the task areas related to hazard risks. As with financial risks, the impact of a malfunction or failure of a safety-critical software-intensive system is on both the owner of the system and those affected by the system, such as internal end users, customers, or the public in general. The main difference in approach between security risk and safety risk evaluations is

Table D.1

Safety-Related Tasks for the Hazard Risk Analysis and Control Phase

Task Areas	Descriptions	Purposes
Identification of potential hazards	Consideration of a full range of types of potential physical and/or environmental harm that could be inflicted if the control system were to malfunction or fail. Such hazards might include the potential release of poisonous or radioactive materials and crashes of vehicles, as well as indirect consequences that result from an event but are not in the zone of impact. Such secondary effects might include the death of a heart-transplant patient because the plane carrying his or her replacement heart crashed or had an emergency landing at another location that is too far away from the patient for the heart to be transported in time.	To develop a list of hazard conditions that could lead to mishaps. These hazards might result from poor or inadequate system design, system malfunction or failure, or intentional or accidental destructive human actions. Also, to determine whether the malfunctioning or failed system can inflict damage to itself so as to inactivate or compromise inherent safety features, where malfunctioning might be the result of an event or inherent system flaw.
Severity estimation	Understanding of the likely severity of each identified hazard. This is often greatly affected by the particular context or environment in which the system operates. For example, damage from the radiation leakage from a nuclear power plant is much greater than radiation leakage from a dentist's x-ray machine.	To determine which hazards might cause the most damage and have the greatest impact, so that extra attention can be given to preventing or avoiding those particular hazards.
Cause identification	Understanding of what might trigger a hazard to actually produce harmful effects. That is to say, the task is to identify the ways in which a system might malfunction (due to design errors, intentional or accidental compromise) or fail that would result in a particular hazard being activated.	To determine the range of possible causes of a hazardous situation related to the compromise, malfunction, or failure of a control system.
Likelihood estimation	Estimation of the probability that a particular hazard type of estimated severity and from a specific cause will result in harm to persons and/or the environment.	To develop estimates of the probabilities that hazards of particular severities will occur due a given cause. These estimates are likely to be derived from past history and predictions from subject-matter experts.
Threat determination	Development of a list of potential attacks on a physically and/or logically isolated safety-critical system.	To identify a list of threats that might become exploits against a safety-critical system by insiders or infiltrators gaining access to the system.
Threat assessment	Understanding the extent of potential damage caused by attacks, which could be experienced by the realization of a threat.	To assess the potential damage that would result from the conversion of a threat into an exploit.

Table D.1 (continued)

Task Areas	Descriptions	Purposes
Exploit determination	Gaining of knowledge as to which exploits have already been used against similar safety-critical software systems, and from that knowledge, extrapolate as to possible future exploits.	To determine those threats that that have been, and could be, realized as exploits against a particular category of safety-critical system.
Exploit assessment	Determination of the technical viability and financial impact—on both the attacker and the victim—if specific exploits were to be activated.	To help decide whether it is worthwhile to plan to protect against specific exploits or if the current status of such a factor as technical viability, say, means that it is highly unlikely to appear.
Weakness determination	Determination of the extent to which safety-critical software systems contain weaknesses. Many of the weaknesses in The MITRE Group's Common Weakness Enumeration (CWE) list are applicable to safety-critical software, although the full list is more suited to security-critical web-facing applications.*	To identify software errors that might make a safety-critical software system vulnerable to compromise.
Weakness assessment	Evaluation of the number and severity of weaknesses exhibited by safety-critical software systems based upon manual and automated program code reviews. Also, an attempt to determine whether particular weaknesses represent real threats to the safety and security of software-intensive systems.	To determine the extent of weaknesses found in the code of safety-critical systems in order to be able to evaluate whether action needs to be taken to spend the effort to correct the weaknesses.
Vulnerability determination	Determination as to whether specific weaknesses represent potential security and safety vulnerabilities which could be exploited.	To arrive at a list of vulnerabilities that result from weaknesses and other factors affecting the security and dependability of safety-critical software.
Vulnerability assessment	Evaluation of the possible impact in the event that a safety-critical software system is attacked successfully via the exploitation of security weaknesses.	To decide whether attacks against known vulnerabilities might succeed, allowing attackers to gain control of the safety-critical software system.
Risk identification	Development of a list of risk categories for both hazards (i.e., how the system might potentially cause harm) and attacks, and listing of risks within those categories.	To identify risks that could affect safety by creating hazardous conditions or could threaten the security of the safety-critical software system so that the risks can be addressed according to their potential impact.

Table D.1 (continued)

Task Areas	Descriptions	Purposes
Risk assessment	Calculation of expected physical, financial and other consequences of system malfunction and failure on the outside world and on the organization that built and/or is running the safety-critical control system. Calculation of the expected losses incurred if the safety-critical software system is compromised by attack from insiders or malicious outsiders who gained access to the system,	To determine the expected losses from the consequences of safety-related malfunctions or failures and/or from security compromises leading to such malfunctions and failures.
Risk mitigation	Reduction or elimination of hazard risks using protective measures (such as containment of dangerous materials, use of protective clothing, seat belts and air bags, etc). Reduction or elimination of risks related to the compromise of the system from intentional or accidental attacks.	To reduce or eliminate the possibility of harm being done by system malfunctions or failures or protect against attacks that might compromise the system.
Risk management	Management and control of the processes relating to identifying, assessing, and mitigating risks related to malfunctioning or failure of safety-critical software systems, and compromise of those systems through successful exploits against vulnerabilities.	To ensure that risks are identified and managed appropriately and that the safety and security of software is maintained at an acceptable level at the least cost.

* *See:* [1].

that when safety is the main criterion, the emphasis is on the harm that could be inflicted on others rather than the financial, reputation, and other losses of the entity running the system, even though the latter could be very substantial. However, we do include in Table D.1 the determination and assessment of threats, exploits, weaknesses, and vulnerabilities, as were incorporated into the Secure SSDLC in Appendix C. Here, we distinguish between hazards, which are those activities than could potentially be done *by* the system, and threats, which are those actions that could be done *to* the system. Up until recently, the latter threat scenarios were largely ignored by software safety engineers, although that is changing with the advent of more cyber-physical systems. Ironically, even isolated safety-critical systems have always been open to compromise by malevolent insiders and by unwitting accomplices, as in the case of the Stuxnet worm. Often, physical controls have been implemented for highly-dangerous control systems (as with the need for two separate and independent operators to activate a nuclear missile launch) but such methods are increasingly being replaced by computer systems.

When considering risks relating to safety-critical software systems, it is important to include both safety and security factors, especially as a security breach can lead to unsafe functioning of safety-critical systems. As mentioned in the main text, safety and security software attributes are frequently dealt with separately in the literature, with little crossover between silos. Only a relatively small number of publications address both safety and security together. Task areas for security-critical software systems for the remainder of the SSDLC were addressed in Appendix C. Many of the task areas for the Safe SSDLC described below are in similar categories to those of the Secure SSDLC, except for the safety orientation of each task area. Certain terms, such as *verification* and *validation*, are more common for safety-critical systems, with *testing* being the term most often used for security-critical systems; the overall concepts are similar. Many of the concepts used are from the JSSSC report [2], which in turn leans heavily on United States military standards. The JSSSC report presents an excellent example of a Plan of Action and Milestones (PAO&M) schedule for developing safe software systems, which also includes consideration of hardware.

In Table D.2, we describe the task areas needed for coming up with generic safety requirements for the software system. As stated in [2]:

> Generic SSRs [software safety requirements] are those design features, design constraints, development processes, "best practices," coding standards and techniques, and other general requirements that are levied on a system containing safety-critical software, regardless of the functionality of the application. The requirements themselves are not ... tied to a specific system hazard ...

Table D.2

Safety-Related Tasks for the Requirements Phase

Task Areas	Descriptions	Purposes
Identification of a complete set of safety requirements	Gathering of safety requirements from generic lists, analysis of system functionality (safety design requirements), causal factor analysis, implementation of hazard controls, as well as stakeholders, including: end users, software safety working group or team, information security team, legal and compliance, internal audit, external auditors, safety and security consultants, human resources, project management office, systems architects, application developers, testers, quality assurance staff, (independent) verification and validation analysts, certification specialists, deployment staff, operators, technical support functions, system administration, and safety and security administrative support functions. Software safety requirements include the list from JSSSC [1], namely, safety design features and constraints, safety coding standards and techniques (e.g., high-integrity programming languages), and safety-oriented development processes.	To ensure that as many safety requirements as possible are gathered from appropriate sources and from analysis of prior safety-critical systems and mishaps, so that the safety requirements will be analyzed, prioritized and included in the overall set of system requirements.
Analysis and prioritization of safety requirements	Prioritization of safety-related requirements by authoritative decision makers such as client organizations, software safety working groups, domain engineers, software quality assurance, software verification and validation, and the like.	To ensure that appropriate safety-related requirements are included in the system architecture and design.
Evaluation of safety risks and consequences	Education of business units so that they understand the risks and consequences of safety-related incidents.	For educational purposes.
Hazard impact analysis and asset protection evaluation	Understanding and determination of the value of various victims and other resources that might be harmed as the result of a safety mishap. Might use values assigned by insurance companies. Determination of the value of assets that need to be protected from security compromises, particularly where security breaches might lead to safety-related mishaps.	To determine the level of prevention that is needed to reduce or avoid harm from mishaps and to choose technical and non-technical safety controls that are appropriate so that they can be incorporated into the safety architecture and design of systems. Also to determine how much protection is needed to avoid compromise of the system that might lead to unauthorized access to sensitive information and/or the taking over of control of safety-critical software systems.

Table D.2 (continued)

Task Areas	Descriptions	Purposes
Legal, regulatory and certification requirements	Review and understanding of legal, regulatory and certification requirements and translation of them into safety policy.	To incorporate relevant laws, regulations and certification requirements into the design development, operations, and disposal of specific safety-critical software-intensive systems.
Agreement upon safety goals	Agreement regarding safety requirements and goals by key decision-makers, including system and safety software engineers, business area management, and senior management, as well as those knowledgeable in legal, regulatory, and certification requirements.	To establish and confirm design goals related to key safety requirements.
Determination of potential threats	Understanding of the sources and forms of threats that might lead to actual harmful events or mishaps.	To be able to assess risks associated with specific hazards and mishaps so that controls to mitigate those risks can be prioritized.
Analysis of potentially damaging activities	Understanding of the range of potential malicious and harmful activities that could affect safety-critical software systems.	To determine activities that might impact the safety of systems so that appropriate controls can be put in place.
Analysis of high-level security vulnerabilities affecting safety-critical software	Understanding of the classes of vulnerabilities and their impact if used to compromise the security of safety-critical software systems.	To be able to address significant vulnerabilities by eliminating (through patching) or bypassing them, and/or by implementing protective capabilities.
Future business goals	Determination as to whether or not the proposed system is extensible so that it can be built to account for likely future safety requirements.	To design the proposed system so as to be able to incorporate anticipated safety requirements in future versions of the software system.
Business/IT operations plans	Understanding of the operational and information technology contexts within which the proposed software system is expected to run—should include full documentation of the responsibilities of each of various participants.	To ensure that the new software system will operate safely in various expected contexts and that it will meet operational safety requirements.

Table D.2 (continued)

Task Areas	Descriptions	Purposes
Deficiencies in current safety program	Review of current safety policy, risk analysis, procedures, standards, processes and technology in order to identify any gaps or deficiencies created by the requirements of the new system. The changing of policy if so dictated by the new software system's requirements. This might occur if the new software system is in a field governed by different regulations and safety certification requirements from previous software developed by the organization.	To understand deficiencies in current safety program in order to address them separately or via the system-level security architecture.
Identifying population of users and its safety requirements	Determination of who will be using the new system and for what purposes. Also, obtaining information about when and for how long users will be expected to use the system, in case there are any time-dependent limitations that might affect system safety.	To understand full set of user safety requirements in order to confirm that the scope and design of the system will satisfy those requirements, and if not, to develop strategies to achieve conformance.
Customer and partner interface requirements	Evaluation of the considerations and constraints that might arise with the new software safety architecture and design for the new software system resulting from new system requirements and system integration. This consideration is particularly important when safety-critical and security-critical software systems are combined to form cyber-physical systems, since the two categories of system are likely to have completely different, and sometimes conflicting, architectures.	To ensure that older and new systems interoperate at the point in time when the systems are interconnected.
Project timeframe	Establishment of an agreed-upon software safety project work plan, including timeframe, milestones and deliverables. Comparison of the software safety work plan, milestones, and deliverables with those of the overall software system development life cycle and resolution of inconsistencies.*	To ensure that there is a basis for acquiring and assigning specific levels of human, technical and support resources, and for applying those resources in order to accomplish given deliverables in time for the assigned milestones. To ensure consistency with the overall work plan.
Prioritized safety solution requirements	Documentation of high-level safety requirements as they relate to users and technology for use in the architecture and design phase.	To ensure that high-level safety requirements are introduced into the development process at an early stage.

Table D.2 (continued)

Task Areas	Descriptions	Purposes
Assessment and verification of software safety standards and criteria (JSSSC [2], pp. 4-69 to 4-71)	Extraction of standards and criteria from general documents, military, government and industry standards and handbooks, project teams' experience, lessons learned from other similar projects, internal documents, and other sources, and verification that the software system is compliant with such standards.	To verify that the software system is developed according to applicable safety-related standards and criteria.
Proposed costs subject to budget constraints	Creation of a software safety budget that covers the development, testing, deployment, integration, operation, decommissioning, and disposal of the system with specific focus on such activities as ongoing safety assessments, vulnerability monitoring, and the like.	To make sure that all relevant and required costs relating to software safety are included within the budget so as to avoid having to go back for additional funding if deficiencies are found.
Approval of safety requirements and associated budget	Explanation to decision-makers and receipt of approval from them for the selected list of safety requirements and respective estimated budgets.	To formalize the safety requirements approval process and obtain written sign-offs for requirements and associated budgets in order to avoid subsequent issues.
Decision as to whether software safety components and systems are to be built internally, bought from a third party or both	Determination as to which software safety tasks and functions can be better assigned to internal staff, third parties, or a combination of resources. This depends on the ability to execute safety architecture, design, testing, verification and validation, implementation, operation, upgrading, replacement, decommissioning, and disposal.	To determine the appropriate assignment of software safety tasks and functions.
Independent safety-oriented review of requirements	Engagement of third parties to perform an assessment of the safety requirements.	To determine if any safety requirements need to be added or modified.

*This is well illustrated in Figure 4.17 on page 4-27 of the JSSSC report [2].

Table D.3

Safety-Related Tasks for the Architecture and Design Phase

Task Areas	Descriptions	Purposes
Create system-level safety architecture	Determination of which safety solutions will be needed to meet the safety requirements and accomplish the goals established in the previous phase.	To document, at a high-level, specific safety requirements and potential technical ways in which those requirements might be met by the system architecture.
Prioritize module safety-criticality analyses	Listing of all modules that involve safety-critical functions and the importance of the module to achieving the required level of safety. Also selecting modules with the greatest safety impact for high priority analysis.	To determine which software modules are safety-critical to assist in deciding upon the priorities of the analyses: to be done by the software safety engineers.
Perform architecture structure analysis (JSSSC [2], pages 4-45 and 4-46)	Construction, at the function level, of the architectural hierarchy in the form of a control tree.	To determine if any safety-related errors or concerns exist in the architecture.
Determine technical security and safety controls	Listing of the requisite technical controls, such as access control, authentication, encryption, etc. to be applied to safety-critical software systems depending upon their current context and future expected contexts (e.g., isolated systems, systems connected to internal networks that are isolated from public networks, systems connected to public networks such as the Internet).	To document technical security controls needed to meet security requirements for safety-critical software systems
Determine nontechnical controls	Listing of the requisite nontechnical controls, such as safety-related manual processes and procedures.	To document those nontechnical controls that are needed to meet predetermined safety requirements.
Perform safety architecture walkthrough	Identification of any deficiencies in the safety architecture so that they might be corrected.	To confirm that proposed technical and nontechnical controls will meet safety-related business requirements effectively.
Create system-level safety design	Determination of specific techniques (i.e., methods and tools) that will be used to achieve the requisite technical and nontechnical controls.	To document specific technical and nontechnical controls that will be used for maintaining the safety of the software system.
Establish top-level technical safety design	Listing of tools and products contributing to the technical solution indicating how they will interoperate with other technical safety elements.	To document the required technical safety elements and how they will be used to achieve various safety requirements.

Table D.3 (continued)

Task Areas	Descriptions	Purposes
Establish top-level nontechnical safety design	Description of safety processes and procedures and their interrelationships, (i.e., safety Concept of Operations (CONOPs)).	To document high-level nontechnical safety processes and procedures.
Analyze costs and benefits for various design elements	Comparison of costs and benefits for various design alternatives and prioritization of these alternatives.	To evaluate the various design scenarios and make economic tradeoffs prior to investing in detailed design.
Detailed design software safety analysis (JSSSC [2], pp. 4-49 to 4-59)	Identification and analysis of those code-level routines, functions, and modules, and related sub-processes, in which software safety requirements should be implemented. Analysis includes consideration of safety interlocks, checks and flags, firewalls, potential hazards due to interfacing, and the like. This is where the cyber-physical interface is examined.	To identify those units (functions, routines, modules) where software safety requirements are to be implemented.
Create detailed technical safety design	Specification of vendor tools, configuration settings, etc. A couple of code-analysis software tools are mentioned in JSSSC [2, pp. 4-59 to 4-60].	To add specific detail to the top-level technical security design.
Create detailed nontechnical safety design	Specification of particular steps in processes and procedures for each process in the CONOP, including specific technical safety standards and configuration procedures.	To add specific detail to the top-level nontechnical safety design
Review technical design at the application and infrastructure levels	Understanding of issues relating to vendor and tool integration and to performance. Performing laboratory testing of safety infrastructure, as needed.	To determine whether any design changes should be made to correct flaws and to prioritize those changes, if any. To make product selections based on architecture requirements.
Creation of secure safety-critical software systems	Education of developers about general and development-related security and safety issues, such as secure development environments and security- and safety-related tools available to developers.	For educational purposes.

Table D.3 (continued)

Task Areas	Descriptions	Purposes
Design training and awareness programs for end-users, operators, administrators, business managers, and executive management	Development of training and awareness documentation and other media (e.g., videos) for various courses depending upon the audience, and establish courses, certifications, and the like. Resources should cover system documentation and operation, help procedures, installation procedures, and legal and regulatory requirements.	To ensure that each area that touches safety-critical software systems has a good understanding of the proper use of security and safety features, activities that should not be done, issues relating to the legal, regulatory and certification requirements, reporting of unusual and potentially infringing and/or dangerous activities, and so on.
Create plan for software safety test phase	Specification of aspects of the software system that are to be tested. These characteristics include safety, security, performance, integrity, availability, resiliency, interoperability, portability, and the like. Functional testing includes assuring that the functional requirements of the software are correct. Functional security testing is used to determine that the software system does not allow unauthorized and unacceptable activities to take place.	To develop a plan for testing functional and nonfunctional features of safety-critical software systems.
Create verification and validation plan	Ensuring that each component of a safety-critical system is built in accordance with safety specifications (verification) and checking that each completed component satisfies its safety specifications (validation), per Braude [3, pp. 26–28].	To develop a plan for the verification and validation of safety-critical software systems
Add or update information security and system safety policies, standards, and/or procedures as necessary	Reevaluation of existing documentation on safety and security policies, standards, procedures, and training in the light of new software systems, particularly those using new, untried and vulnerable technologies.	To ensure that all safety-related documentation, including that which relates to security and safety, has been brought current with respect to new software systems.
Create plan for mitigation of key application and infrastructure hazards	Population of a database of hazards obtained from various sources such as software and hardware product vendors, safety-related websites, and so on, Development of internal procedures for applying fixes and monitoring and reporting the status of fixes against the entire population of systems that might be affected by or create hazards.	To document hazards for applications, middleware, system software and infrastructure, both internally-developed and purchased off-the-shelf. To prioritize available fixes (patches) and ensure that those representing the greatest risk are applied timely.

Table D.3 (continued)

Task Areas	Descriptions	Purposes
Determine security and safety requirements for development, test and operational environments	Determination of the processes, procedures and technologies that assure a reasonable and suitable level of security and safety. Examples include version control, access controls, segregating development, testing and production environments, scrubbing data used in the development and testing phases so as not to include sensitive information. Note that the security of the processes for creating safety-critical software is required to mitigate nefarious activities during the SSDLC. This is different from the ability of the final system to withstand damaging attacks, which is intrinsic to the design of the system.	To ensure that, as the software development passes through its various phases, the level of security is high enough so that the software will not be compromised in its travel by such dangers as external hacker attacks, and insider-instigated risks.
Independent safety-oriented review of architecture and design of software system	Engagement of third parties to perform a safety assessment of the architecture and design of the software system.	To determine if there are any safety aspects of the architecture and design that need to be added or modified.

According to the JSSSC report [2], the requirements are based up experience of errors or omissions that resulted in mishaps or potential mishaps.

When the above tasks have been completed, the project team should have a complete set of approved safety requirements for the proposed software system, along with formal approvals of the requirements, funding, and resource-levels for the implementation of those requirements. The requirements are then used in the subsequent software system safety architecture and design phase, which is described in Table D.3.

Once the safety architecture and design tasks have been accomplished, the results need to be incorporated into the application development process as indicated in Table D.4. As noted in Table D.3, it is important to secure the phases of the development life cycle for safety-critical software systems and to ensure that appropriate security services, such as strong authentication, restrictive authorization, and strong encryption, are incorporated into safety-critical systems and that the monitoring, reporting of events, and incident response are built into safety-critical software systems even though they are not expected to be exposed to external attacks.

Verification tasks are generally performed at each phase throughout the Safe SSDLC, with validation conducted when the software has been developed, and certification applied for when the fully-integrated software system is finalized. These tasks are shown in Table D.5.

It would appear that too little attention is being paid to the deployment phase in many projects, yet this phase subjects the organization to huge potential risks, particularly if the deployment involves integration with other software systems that are not under the direct control of the deploying entity. A case in point is the faulty deployment of a mission-critical trading system on August 1, 2012 by the Knight Capital Group. The launch of the updated system resulted in a catastrophic malfunctioning and an immediate financial loss of $440 million [4]. While this example does not have direct safety implications, it is indicative of the type of disaster that can occur during the deployment phase, and it provides strong evidence that supports paying full attention during this critical phase.

There are two types of safety risk during deployment. One is that the deployment process might be deficient in that it does not adequately check that all the modules needed are being installed and that no superfluous modules remain in the software system. The other relates to intrinsic problems within the software itself due to such causes as inadequate requirements, specifications, and testing in the context into which the software is being released.

The task areas relating to the deployment of safety-critical software systems are described in Table D.6.

Table D.4
Safety-Related Tasks for the Development Phase

Task Areas	Descriptions	Purposes
Set up secure development environment	Implementation of processes, procedures and technology needed to create a secure development environment.	To ensure that access and use of source code, component configurations, documentation, and the like are controlled and monitored, that deviations are reported, and that responses to incidents are preplanned.
Security-related training for teams developing safety-critical systems	Provision of training with respect to safe and secure coding practices, including choice of programming languages, platforms, and infrastructures most appropriate for safety-critical systems. Also making developers aware of general safety and security issues, other development-related security issues (e.g., secure development environment, see above), and safety-related and security-related development and testing tools.	To ensure that developers are adequately trained in secure coding practices, as well as being made aware of other relevant security-related matters, and are familiar with particular technologies that support safety-critical software development.
Security-related and safety-related training for infrastructure teams	Provision of training with respect to the configuration and installation of middleware and other system software and services, as they relate to safety and security.	To ensure that infrastructure teams are adequately trained in the safety and security aspects of configuring and operating platforms and infrastructure upon which the developed software will run.
Develop application-level safety components	Creation and/or acquisition of safety-related components and services, which also satisfy security requirements for safety-critical software systems.	To develop and/or acquire components that meet safety-related and security-related requirements and specifications.
Implement safe and secure infrastructure in support of application software	Establishment of a safe and secure infrastructure and environment in support of application development and testing activities.	To provide a safe and secure environment for developers and testers activities.
Set up hazard, vulnerability, and mitigation tracking process	Creation of a hazard, vulnerability, and mitigation tracking database, possibly integrated with a quality assurance or error tracking system, with the ability to record, prioritize, sort, monitor, report, and indicate action taken to reduce the risk posed by hazards and vulnerabilities.	To ensure that hazards, vulnerabilities, and mitigation efforts are identified, recorded, prioritized, resolved, and marked as completed.

Table D.4 (continued)

Task Areas	Descriptions	Purposes
Plan safety testing (JSSC [2, pp. 4-63 to 4-65])*	Development of safety test plans for prospective, current and scheduled future versions of the software system. The plans should include testing of software requirements and performance-related requirements.	To ensure that adequate safety-related testing is performed.
Unit, integration and regression testing of safety components	Development of plans for testing that safety components have been correctly integrated into the overall safety-critical system.	To ensure that safety-related components, created and/or used by developers, are working properly.
Independent review of development processes and products	Engagement of third parties to perform an independent assessment of the development processes and procedures and their conduciveness to developing safe software.	To determine if there are any safety issues, relating to the development of the software system, which need to be added, deleted, or otherwise modified.

*It should be noted that, in the JSSC report [3], the terms *testing* and *verification and validation* appear to be used interchangeably. In this book, *testing* is considered to be one component of both verification and validation. The latter activities include such tasks as the inspection of requirements and specifications and how they have been implemented, determining whether requirements have been satisfied, and obtaining management signoffs and customer approvals.

Table D.5
Safety-Related Tasks for the Verification, Validation, and Certification Phase

Tasks	Subtasks	Purpose
Code review	Review of application code to verify that it meets safety requirements and specifications.	To determine if there are any code-level safety issues and, if so, to address them.
Testing of configuration procedures	Review of configuration procedures and verification that they are usable by administrators for setting up networks and system components to meet safety standards.	To determine if there are any procedural issues in need of resolution.
System testing	Testing to determine whether the technical safety components of the software system are functioning together and conform to architecture and design specifications.	To determine if there are any system-level and system-integration issues in need of resolution.
Performance and load testing	Testing to determine whether system performance and production load requirements meet prespecified levels with safety components and features operating.	To determine if there are any safety-related issues relating to performance that need to be resolved.
Usability testing	Perform tests to determine whether system meets end-user usability and ergonomic needs.	To determine if there are any usability issues, which could affect safety, in need of resolution.
IV&V (independent verification and validation) assessments	Engagement of third parties to perform verification and validation assessments in order to determine whether there are any application, platform, and/or infrastructure safety issues.	To determine if there are any safety issues that need to be addressed.
Certification	Obtaining safety certification at a particular level in order that the software system can be used for its intended purpose.	To obtain certification so that the software system can be deployed as required.

Table D.6

Safety-Related Tasks for the Deployment Phase

Task Areas	Descriptions	Purposes
Pilot deployment	Identification of any remaining issues related to all areas of software system planning, design, architecture, development, testing, implementation, and decommissioning.	To allow development and architecture teams to make adjustments to operational processes, systems architecture, design, and programs prior to release.
Transition from final testing to operations	Ensuring that the final system, just prior to deployment, addresses adequate levels of integrity and that appropriate change management procedures, encryption of transmissions, interim access authorizations, and so on are in place. Also assuring that all required processes have been reviewed and tested.	To ensure that the transition process conforms to good safety and security practices.
Comparison of system files	In the case of a replacement system, making sure that system files of the system that is about to be launched are compatible to the original system files and that they match, thereby assuring the authenticity of the files.	To ensure the authenticity of system files to be moved into the operational environment.
Comparison of data files	In the case of a replacement system, making sure that data files of the system that is about to be launched are compatible with the original data files and that they match, thereby assuring the authenticity of the data.	To ensure the authenticity of data files to be moved into the operations environment.
Training and awareness	Ensuring that all appropriate stakeholders (e.g., end users, administrative personnel, technical support, and operations staff) are trained and are aware of the safety functions and attributes of the system.	To ensure that those who will use and support the new system are sufficiently familiar with the safety requirements of the system in order to perform their respective functions.
Full-scale deployment	Making sure that the existing infrastructure meets safety requirements prior to installation of the new system. Checking that all safety-related procedures have been established and then followed throughout the deployment phase.	To avoid deploying the system into a hazardous environment.
Independent safety-oriented review of deployment	Engaging third parties to perform safety assessments during the deployment, predeployment, and post-deployment.	To determine whether there are any safety-related deployment issues, and which were missed originally but that need to be resolved.

Table D.7
Safety-Related Tasks for the Operations/Maintenance/Support Phase

Task Areas	Descriptions	Purposes
Testing and migrating to new software version	Analysis of new versions of software to determine which hazardous conditions may have been resolved and which hazardous conditions still exist.	To determine the hazards of current and replacement software versions in order to arrive at a revised list of hazards to be resolved.
Risk reviews	Performing a new risk assessment if customer risk conditions have materially changed due to the installation of the new system.	To determine whether safety risks have increased with the new version and therefore need to be mitigated.
Safety awareness program	Maintain a safety awareness program to take into account changes in threat and vulnerability environments, the addition of new customers, employees, and consultants, and changes in the laws, regulations, and certification requirements.	To keep security awareness and training programs up-to-date with respect to content and the population to be trained.
Hazard and vulnerability risk assessments	Perform periodic hazard and vulnerability risk assessments to account for changes in systems, technologies, and personnel.	To ensure that new and changing vulnerabilities are recognized and that a program for their resolution is enacted.
Auditing, logging, monitoring, archiving	Implement appropriate logging, archive procedures, and review the logs periodically in order to detect any questionable safety-related activities.	To determine whether any anomalous activities are taking place and, if so, to take action to halt such activities, prevent future activities, and take appropriate steps against perpetrators.

Once the software system has been deployed, there are ongoing operational, maintenance, and support functions that are required in order to keep the system running in a safe manner, as described in Table D.7.

As long as there are few changes to the system and its environment, it can be expected that the system will remain relatively stable over long periods of time. Any changes can adversely affect safety, and so it is appropriate to conduct periodic safety reviews that will consider the safety impact of any such changes. However, in many cases, the next significant period begins when there is an impetus to close down the system, whether such pressure to change systems is internally or externally initiated. For example, at some point, safety-critical systems will become obsolete and will need to be replaced within the relatively near future, say 18 months to two years. This will result in the need to discontinue the current system and dispose of any traces of the applications. For safety-critical systems, it is very important that the disposal of the system being replaced is done in a safe and secure way. The tasks relating to this decommissioning and disposal phase are described in Table D.8.

Levels of Involvement

In Table D.9, we show the overall involvement of the various groups and teams in the safety aspects of the life cycle. The entries in this table are normative. As expected, the main participants are the safety, security and audit team members, with the other areas generally moderately involved. This differs from Table C.17 in Appendix C, mainly with respect to Table D.9 showing relatively lower involvement of the application development and infrastructure teams and business managers for safety-critical systems. This is attributable to safety being more oriented towards the physical world, resulting in somewhat less involvement by the more technical teams.

A Note on Sources

The tables in this appendix were developed in large part from the JSSC report [2]. The structure and many of the entries in the tables were derived from McBride [5], as was the case for Appendix C. However, as in Appendix C, the phases relating to risk analysis and control, decommissioning, and disposal were added.

Table D.8
Safety-Related Tasks for the Decommissioning and Disposal Phase

Task Areas	Descriptions	Purposes
Integrity of the old and new systems during transition of the old system to a decommissioned state	Ensuring that adequate plans are in place for decommissioning and disposing of the software and any media or hardware on which the system or its components depend.	To avoid problems with incompatibilities, data destruction or loss, inappropriate authentication and authorization, and the like.
Transition from operational to decommissioned state	Ensuring that the new replacement system can operate on its own with respect to safety functions prior to the decommissioning of the old system and that operators can switch back to the old system (instantly for real-time systems) in the event that dangerous behavior is observed during the initial period of operation when the new system takes over. Also making sure in advance that needed information is will be collected and reported, so that action can be taken to minimize the impact of a malfunction or failure.	To avoid the possibility of neither the old system nor the new system being available for any period of time. Also to ensure that the safety functions of the system remain effective at all times prior, during, and after the transition.
Comparison of system files	Before the old system is decommissioned, making sure that the system files of the new system are compared to the original system files and that they match.	To ensure the authenticity of system files that are to be used by the new system before the old system has been decommissioned.
Disposal of application and system files	Ensuring that all valuable application and system files that are no longer needed are either removed from magnetic media or that the media on which the data resides, including paper hard copies, are physically destroyed.	To ensure that the intellectual property contained in the application and system files will not fall into the possession of those unauthorized to access the application and system files, especially those who might use the information for nefarious purposes.
Comparison of data files	Making sure that the data files of the new system, which is about to replace the old system, which in turn is to be decommissioned, are compatible with the original data files and that they match, thereby assuring the authenticity of the data.	To ensure the authenticity of data files, which are to be used by the new system. Where possible, to ensure that the data files are backward compatible, meaning that the new system can still read and operate upon the data files created and used by the old system.

Table D.8 (continued)

Task Areas	Descriptions	Purposes
Disposal of data files and media	Ensuring that all sensitive data that is no longer needed is either removed from magnetic media or that the media on which the data resides, including paper hard copies, are physically destroyed.	To ensure that the data will not be obtained by those unauthorized to access the data, especially those who might use the data for fraudulent or other nefarious activities.
Disposal of hazardous materials and equipment	Particularly for embedded software, making sure that any hazardous materials or dangerous equipment are suitable disposed of or destroyed.	To ensure that the disposal process does not cause harm.
Training and awareness	Ensuring that all appropriate stakeholders (i.e., end users, administrative personnel, technical support, and operations staff) are trained and aware of the safety functions and attributes of the new system and have in hand a set of procedures for the transition.	To ensure that those who will use and support the new system are sufficiently familiar with the security aspects of the new system versus the old system as required to perform their respective functions.
Full-scale decommissioning	Ensuring that the existing infrastructure is safe and secure prior to transitioning from the old system to new system, possibly including a period of parallel operation.	To avoid creating a hazardous or vulnerable environment during the decommissioning of the old system and installation of the new system.
Independent safety-oriented review of decommissioning and disposal	Engaging third parties to perform safety assessments of the transition pre- and post-decommissioning.	To determine whether there are any safety-related issues relating to decommissioning and disposal that need to be resolved.

Table D.9
Summary of Levels of Overall Involvement by Phase

Phase	Overall Level of Team/Group Involvement								
	Safety Team	Security and Risk Team	Application Development Team	Infrastructure Team	Project & Change Managers	Business Unit Managers	Operations Managers	Internal & External Auditors	Executive Management
Requirements	High	Moderate to high	Low to moderate	Low to moderate	Low to moderate	High	Low to moderate	Moderate to high	Moderate to high
Architecture and Design	Moderate to High	Moderate	Low to moderate	Low to moderate	Low to moderate	Moderate to high	Moderate to high	High	Moderate to high
Development	High	High	High	High	High	Moderate	Moderate	High	Low to moderate
Verification, Validation and Certification	High	Low to moderate	Moderate to high	Moderate to high	Moderate to high	Moderate	Low	High	Low
Deployment	High	Moderate to high	Moderate to high	Moderate to high	High	Moderate	Moderate	High	Moderate
Operations, Maintenance and Support	High	High	Moderate	Moderate	Moderate	Moderate to high	Moderate	High	Low to moderate
Decommissioning and Disposal	High	Moderate to high	Low	Low	Low	Low	High	High	Low
Overall Life Cycle	High	Moderate to high	Moderate	Moderate	Moderate	Moderate	Moderate	High	Low to moderate

Endnotes

[1] *CWE Version 2.2,* 2012, S. M. Christey, J. E. Kenderdine, J. M. Mazella and B. Miles (eds.), and R. A. Martin (project lead), available at http://cwe.mitre.org/data/published/cwe_v2.2.pdf, last accessed on August 12, 2012.

[2] Joint Software System Safety Committee (JSSSC), *Software System Safety Handbook: Technical & Management Team Approach,* 1999. Available at http://www.system-safety.org/Documents/Software_System_Safety_Handbook.pdf. AccessedJuly 23, 2012.

[3] Braude, E. J., and M. E. Bernstein, *Software Engineering: Modern Approaches, Second Edition,* Hoboken, NJ: John Wiley & Sons, 2011.

[4] See, for example, http://dealbook.nytimes.com/2012/08/02/knight-capital-says-trading-mishap-cost-it-440-million/, last accessed on August 12, 2012.

[5] McBride, P., and E. P. Moser, *Secure System Development Life Cycle (SDLC) Building Security into Your System—Not Bolting It On After the Damage is Done,* Atlanta, GA: METASeS, 2000. Available at http://lazarusalliance.com/horsewiki/images/3/38/Secure-System-Development-Life-Cycle.pdf. Accessed August 5, 2012.

About the Author

C. Warren Axelrod is a leading authority on information technology, security, and privacy for the financial services industry. He has held executive positions at such institutions as SIAC, HSBC Securities and Pershing. Most recently, he was the chief privacy officer and business information security officer for the U.S. Trust.

He focuses on cyber security, risk management, and critical infrastructure protection, predominantly in the financial services industry, as well as at the national level. His recent consulting work has involved him with safety-critical control systems. He represented the banking and finance sector at the Y2K command center in Washington, D.C. (December 1999–January 2000). He testified at a congressional hearing on cyber security (November 2001) and is active in the Cloud Security Alliance.

In 2009, he received the Michael P. Cangemi Best Book/Best Article Award presented by ISACA for "Accounting for Value and Uncertainty in Security Metrics," published in the *ISACA Journal*, in 2008. He was honored with the prestigious Information Security Executive (ISE) Luminary Leadership Award in 2007. He received a Computerworld Premier 100 IT Leaders Award and Best in Class award in 2003.

Dr. Axelrod has written several books, including *Outsourcing Information Security* (Artech House, 2004). He was coordinating editor of *Enterprise Information Security and Privacy* (Artech House, 2009). He holds a Ph.D. in managerial economics from Cornell University and a B.Sc. in electrical engineering and an M.A. in economics and statistics from Glasgow University.

He is certified as a CISSP and CISM.

Index

Multicast and Group Security, Thomas Hardjono and Lakshminath R. Dondeti

Non-repudiation in Electronic Commerce, Jianying Zhou

Outsourcing Information Security, C. Warren Axelrod

Privacy Protection and Computer Forensics, Second Edition, Michael A. Caloyannides

Role-Based Access Control, Second Edition, David F. Ferraiolo, D. Richard Kuhn, and Ramaswamy Chandramouli

Secure Messaging with PGP and S/MIME, Rolf Oppliger

Securing Information and Communications Systems: Principles, Technologies and Applications, Javier Lopez, Steven Furnell, Sokratis Katsikas, and Ahmed Patel

Security Fundamentals for E-Commerce, Vesna Hassler

Security Technologies for the World Wide Web, Second Edition, Rolf Oppliger

Techniques and Applications of Digital Watermarking and Content Protection, Michael Arnold, Martin Schmucker, and Stephen D. Wolthusen

User's Guide to Cryptography and Standards, Alexander W. Dent and Chris J. Mitchell

For further information on these and other Artech House titles, including previously considered out-of-print books now available through our In-Print-Forever® (IPF®) program, contact:

Artech House	Artech House
685 Canton Street	16 Sussex Street
Norwood, MA 02062	London SW1V HRW UK
Phone: 781-769-9750	Phone: +44 (0)20 7596-8750
Fax: 781-769-6334	Fax: +44 (0)20 7630-0166
e-mail: artech@artechhouse.com	e-mail: artech-uk@artechhouse.com

Find us on the World Wide Web at: www.artechhouse.com